PUSWHISPERER 4

PUSWHISPERER 4:

A FOURTH YEAR IN THE LIFE OF AN INFECTIOUS DISEASE DOCTOR

by Mark Crislip, MD

Bitingduck Press
Altadena, CA

Published by Bitingduck Press
ISBN 978-1-68553-019-8

For information contact
Bitingduck Press, LLC
Altadena, CA
notifications@bitingduckpress.com
http://www.bitingduckpress.com
Cover images from Pixabay: bacteria by Raman Oza, human figure by Gerd Altmann. Used under terms of Pixabay license.

Disclaimer

Adapted from Medscape blog entries, covering cases from September 2011 to Septemper 2012.

The information here is not meant to diagnose, treat, cure or prevent any disease and is not intended for self-diagnosis or self-treatment of medical conditions that should be managed by a qualified health care provider.

To protect patient confidentiality, all demographic and identifying patient information was changed.

To my eldest, Jeffery Donald Crislip, who knows where, all, the commas, belong.

Summertime Is Encephalopathy Time

ANOTHER young person with fevers, altered mental status, sore throat, and a rash. She had recently been hiking in the Cascades Mountains in western North America and now has a new partner. The rest of the ID (infectious disease) history is negative.

An exam shows a mildly confused female with a rash that spares the palms and soles. Labs have normal WBC (white blood cells), a mild lymphopenia, and the spinal tap has a protein of 125, a normal glucose, and 22 WBCs, mostly lymphocytes. This must be the fifth similar presentation so far this summer for this patient, and there is the same differential diagnosis: enterovirus (unlikely with encephalopathy), West Nile (still hasn't hit the Pacific Northwest; I think the beer is protective), syphilis, HSV (herpes simplex virus)(but with a normal MRI, which doesn't fit) and all of the usual causes of aseptic meningitis that should not be causing encephalitis.

This time? An HIV (human immunodeficiency virus) test is negative, but the HIV viral load is 1,500,000 copies per milliliter and the CD4 T-cell count is 421, which is low. So acute HIV it is.

The last time I saw an acute meningoencephalitis as a manifestation of acute HIV was as a fellow. The patient had had a single unprotected sexual contact in Mexico, then presented nine days later just like this patient. I have looked many times since, but had not made the diagnosis again until today. There are only nine PubMed references on the topic, so it is rarer than I had thought. The problem with acute HIV infection is that it is indistinguishable from the more common causes of headaches, sore throat, fevers, and mild adenopathy, which are common in the clinic. Unless the patient volunteers or is asked about risk factors, it will likely be called a viral syndrome. I am generally suspicious, figuring that everyone lies about sex, drugs, and rock 'n' roll, and as far as I am concerned, the only risk factor to consider in HIV is whether the patient is human. I wonder if the neurological involvement is more the virus or the host. There is a lot of information on the pathophysiology of the various HIV neurological complications, but none that relate to meningitis. I bet there is some receptor polymorphism that leads to this presentation.

And should acute infection be treated, and if so, for how long? There are no randomized controlled trials that really look at the issue, but observational studies suggest that there may be some long-term immunologic benefits:

> *"Initiation of HAART within 2 weeks of antibody seroconversion was associated with viral load and CD4+ T cell count benefits for 24 weeks after termination of HAART, with there being trends toward a longer-term benefit. Later initiation of HAART was associated with a persistent but decreasing CD4+ T cell count benefit and a loss of the viral load benefit by week 72 after discontinuation of treatment."*

and

> *"Treatment during PHI did not lower viral set point. However, patients treated during seroconversion had an increase in CD4 lymphocytes, whereas untreated patients experienced a decrease in CD4 lymphocytes. Time until reaching CD4 lymphocytes <350/ microl was significantly shorter in untreated than in treated patients including patients with CD4 lymphocytes <500/ microl during seroconversion."*

It has been years since I have seen an acute HIV infection, and there is no local or interweb consensus. I am inclined toward suggesting therapy, but although much improved, the patient is still a bit too confused to participate in the decision.

Rationalization

del Saz, S.V. et al. Acute meningoencephalitis due to human immunodeficiency virus type 1 infection in 13 patients: clinical description and follow-up. Journal of neurovirology 14, 474-479 (2008).

Dunfee, R. et al. Mechanisms of HIV-1 neurotropism. Current HIV research 4, 267-278 (2006).

Hecht, F.M. et al. A multicenter observational study of the potential benefits of initiating combination antiretroviral therapy during acute HIV infection. The Journal of infectious diseases 194, 725-733 (2006). Koegl, C. et al. Treatment during primary HIV infection does not lower viral set point but improves CD4 lymphocytes in an observational cohort. European journal of medical research 14, 277-283 (2009).

It Is What I Do Best

THE patient is an elderly female admitted with fevers for eight days. Workup is negative and they call me.I go through my usual aich and pea, review the labs and x- rays, and come up with nothing. She has been nowhere, has done nothing. CBC (complete blood count) chemistry panel, urinalysis, and chest x-ray are all unrevealing. No reason to be found for her fevers of about a week's duration.Once a patient is in the clutches of the medical industrial complex, tests must be ordered. The patient can't sit there, febrile, without looking for a reason. Don't stand there; do something.The longer I am in medicine, the more my motto is, "Don't just do something; stand there." One of two things happen if you watch and wait: either the diagnosis will become obvious even to me, or, more likely, the patient will get better. It is cases like this where I quote the Chinese aphorism:"Muddy water. Let stand. Will clear."

So we did nothing, and within forty-eight hours, the fevers were gone. I took full credit.Older people, as a rule, do not have trivial causes of their fevers. Once you are in your sixties, you have been exposed to most of the viruses that cause viral syndromes, so anyone over age fifty-five (I will be elderly in just a year) who presents to the emergency room febrile deserves a careful evaluation for a serious, treatable cause of the fevers.

"Elderly patients fail to manifest identifiable clinical features indicative of bloodstream infection. The sensitivity and specificity of the best statistical model for identifying bacteremic elderly patients suggest that clinical indicators alone are unreliable predictors of bacteremia in the geriatric ED population studied."

But once the basics are excluded (pneumonia, urinary tract infection, and cellulitis), the aich and pea is not suggestive, and the screening labs are nonfocal; all you can do is wait. There is a reason why you do not embark on the classic fever of unknown origin workup until the patient has had a fever for at least 14 days: Fevers almost always go away on their own.Early in my career, against my better judgment (the patient was the husband of a nurse), I started the FUO evaluation on day 10. By day 14, the fevers were gone.Of course, it takes time and experience to learn the importance of doing nothing, and not every patient is happy when you tell them that yes, you have a fever, and that I am going to watch it and do nothing. If you explain the rationale, most of the time they agree, but on occasion I have had that look from the patient, or from a family member, that says, "You are the biggest dumbass I have ever seen in my life." Probably true. But I often know when the best course of action is inaction. The *New England Journal of Medicine* had a commentary recently entitled,"The Art of Doing Nothing." It is what I do best, and my wife can attest to that.

Rationalization

Rosenbaum, L. The art of doing nothing. The New England journal of medicine 365, 782-785 (2011).

Fontanarosa, P.B., Kaeberlein, F.J., Gerson, L.W. & Thomson, R.B. Difficulty in predicting bacteremia in elderly emergency patients. Annals of emergency medicine 21, 842-848 (1992).

Poll Results

When in doubt,

- Google. 46%
- get a consult. 11%
- quote an aphorism. 18%
- pretend you aren't. 22%

More Rash Decisions

I SAW two patients with extremely bad lymphedema: one because of lymph node dissection years ago, and the other due to obesity. With lymphedema comes recurrent cellulitis and the occasional bacteremia, and both of these patients are admitted with yet another case of cellulitis.

Recurrent cellulitis with lymphedema is always due to Group A *Streptococcus* aka *S. pyogenes*. Well, most always. Maybe 99.99999% of the time. You don't have to worry about MRSA(methicillin -resistant Staph aureus)or *Pseudomonas* or any of the other beasts that create fear, uncertainty, and doubt and lead to the occasional somewhat excessive course of antibiotics.

All they need is penicillin or amoxicillin.

One patient had, as is often the case with recurrent Group A Streptococcal cellulitis, groin pain preceding the cellulitis by about two hours. I have mentioned before that groin pain is a not uncommon symptom of cellulitis, especially with what I think is Group A streptococcus, erysipelas, of the leg.

As if I can tell the microbiology of cellulitis just by looking at it. The literature suggests that health care workers can't tell the cause of cellulitis by looking at it, but then they never included me in the study. Sentinel groin pain has not, as best as I can determine, been described in either journals or in textbooks. The patient was on piperacillin/tazobactam and vancomycin. Aren't they always? Some days it seems my main job is to stop piperacillin/tazobactam and vancomycin. I changed him to amoxicillin, dosed for his increased volume of distribution.

The other patient was on vancomycin, despite positive blood cultures for Group A streptococcus, because he was allergic to penicillins. Or was he?

What was the allergy? As a child he had a maculopapular rash to amoxicillin given while he probably had a case of mono.

> *"70–100% of patients receiving amino-penicillins during a florid Epstein–Barr virus (EBV)-infection develop a maculopapular rash."*

The monospot has a sensitivity of 80 to 97%, so the rash to amoxicillin may be a bit more sensitive, but is probably less specific.

As a rule, the maculopapular rash from amoxicillin can be ignored as predictive of a type 1 reaction, and penicillins may be given later with impunity—at least that is what I was always told. There does not seem to be a huge database to support the assertion:

> *"This increased sensitivity to ampicillin in the presence of viral infection was shown to be a transient phenomenon in that only 2 of 20 patients experiencing an ampicillin-induced rash during infectious mononucleosis exhibited a recurrent rash when receiving the drug after resolution of the infection."*

Be that as it may, there is no better drug for *S. pyogenes* than penicillin, so I challenged him with amoxicillin.

I have a wee bit of superstition. When I have a patient with a retropharangeal abscess or similar infection, I always have a tracheostomy kit in the room. It guarantees it will never be needed. And whenever I challenge an alleged penicillin allergy, I always have an EpiPen in the room. It has never been needed.

The patient took the amoxicillin just fine, and was sent home on it.

Rationalization

Jappe, U. Amoxicillin-induced exanthema in patients with infectious mononucleosis: allergy or transient immunostimulation? Allergy 62, 1474-1475 (2007).

Nazareth, I., Mortimer, P. & McKendrick, G.D. Ampicillin sensitivity in infectious mononucleosis—temporary or permanent? Scandinavian journal of infectious diseases 4, 229-230 (1972).

Svensson, C.K., Cowen, E.W. & Gaspari, A.A. Cutaneous drug reactions. Pharmacological reviews 53, 357-379 (2001).

Blindsided

Most of the time a disease in my differential diagnosis turns out to be the final diagnosis, if a diagnosis is made. Dr. House usually has the answer after fifty minutes, but I am not so fortunate. More often than not, I do not even get a firm diagnosis, just a presumptive one, but as long as the patient gets better, I am happy not to know. Well not happy. But not unhappy. And on occasion the diagnosis comes from so far afield, I can only shake my head and mutter.

The patient is a middle-aged white male. No history, no exposures, no nothing. His life is every bit as uneventful as my own.

Shortly after a trip to Sacramento, he develops the worst headache of his life, bilateral infiltrates with a nonproductive cough (a cough-till-you-puke cough), and, on CT(computerized tomography), big hilar lymph nodes.

He gets the four-star workup and everything, and I mean everything, is negative: no *Coccidioidomycosis*, negative spinal tap, negative serologies for everything he could have and could not have, and two negative bronchoscopies. He remains febrile, with a bad headache and coughing, and then slooooowly gets better over two weeks. The improvement is probably in spite of anything we do, rather than due to anything we do.

Eventually, it is decided to go after the lymph nodes in the chest, and for that he needs to be transferred to another hospital.

And the lymph node shows?

"Lymph nodes with near-complete replacement by noncaseating epithelioid granulomas, consistent with Boeck's sarcoid."

Cultures and special strains are all negative.

Didn't see that coming.

This is acute sarcoid, which has to be an infectious disease, but due to what? The clinical course and pathology suggest an infection and he was improving on his own, although sarcoid may be due to issues with the patient's ability to respond to infection more than the infection itself.

> *"Is CCR-2 an appropriate gene to study in sarcoidosis? Candidate genes may be identified by genome-wide screens using linkage analysis or by a functional approach, where a specific gene is chosen based on its likely contribution to disease pathogenesis. Both approaches support the choice of CCR-2 as a candidate gene in sarcoidosis. A recent study, reporting a genome-wide search for genes predisposing to sarcoidosis, involved 63 families with two or more siblings diagnosed with sarcoidosis (8). This study confirmed a prominent association of sarcoidosis with the major histocompatibility locus on chromosome 6 (the site of HLA class II and TNF genes). It also revealed an association with the short arm of chromosome 3 where the CCR-2 gene is located. CCR-2 is a chemokine receptor for the macrophage chemotactic proteins CCL-2, 8, 12, and 13. CCR-2 is expressed predominantly on macrophages, monocytes, dendritic cells, and T cells, and regulates inflammatory cell recruitment and function at sites of inflammatory responses (9). CCR-2 and its ligands have been shown to regulate cellular accumulation during granuloma formation in mice (10)."*

Acute sarcoid is a rare presentation of the disease, and usually with an acute arthritis syndrome that mimics Still's disease—but he had no joint complaints at all. And it can present as a meningitis, although his only abnormality was a few white blood cells in his cerebrospinal fluid. Hardly an impressive lab finding.

As he improved on his own, he never had a slug of steroids, which is the usual treatment.

It seems that when I get blindsided by a disease, it is always a rheumatologic process, if sarcoid falls into that category. I probably will never see another like it—although the residents will hear about it for the rest of my career.

Rationalization

Case records of the Massachusetts General Hospital. Weekly clinicopathological exercises. Case 4-2001. A 26-year-old man with pain, erythema, and swelling of the legs. The New England journal of medicine 344, 443-449 (2001).

O'Regan, A.W. & Berman, J.S. The gene for acute sarcoidosis? American journal of respiratory and critical care medicine 168, 1142-1143 (2003).

Teh, L.S. et al. Acute sarcoidosis: a difficult diagnosis. Rheumatology 39, 683-685 (2000).

Poll Results

When I get fooled it is

- an infection. 0%
- a rheumatologic diagnosis. 41%
- a cancer. 11%
- I am never fooled. 7%
- There's an old saying in Tennessee — "I know it's in Texas, probably in Tennessee — that says, fool me once, shame on you, shame on you. Fool me, you can't get fooled again." 37%
- Other Answers 4%
  -I often fool myself.

A Swell Joint

THE Great Pacific Northwest is the only place to live, but it has a few downsides. As an ID doc, I will admit that Oregon is a touch on the dull side. Very few endemic illnesses. A little relapsing fever in eastern Oregon and a growing number of *Cryptococcus gattii* infections. Not much Lyme, which is good (a few chronic Lyme patients who don't have chronic Lyme) and bad (I only see a real Lyme every couple of years). My livelihood does depend on people having horrible luck and getting weird infections. No one is going to call me to see an outpatient UTI or pneumonia.

The patient is traveling on business to the East Coast, and while there, he has the abrupt onset of a swollen knee. It is tapped, and not particularly purulent, so it is treated with NSAIDS (nonsteroidal anti-inflammatory drugs) and he returns to Oregon.

It is a good thing he had the tap in New York, as they sent the fluid for a *Borrelia* polymerase chain reaction (PCR) test, and a few days later, the patient was called to say it was positive. PCR for *Borrelia* would probably not have been sent if he had presented in Oregon, because we have no Lyme worth writing about. Unless, of course, he saw an ID doc who would take an exposure history.

The patient saw an orthopedist, who then called me. I have a case of Lyme; what do I do?

Most of the time in Oregon, Lyme is based on serology sent by naturopaths to labs whose provenance I question. So I asked how it was diagnosed, and he said PCR on the knee fluid.

Huh. I have never had the need to order a Lyme PCR. How good is it?

There are over 10,000 PubMed articles each year concerning infection, and I probably read 1% of them. Part of writing this book, besides the fame and wealth—well, the fame—well—is I do it to keep up. Writing keeps me looking things up, and there is no better way to imprint information on my brain than to write on a topic. It is a good chance to brush up on Lyme diagnostics.

So I find that the reference lab we use, ARUP (not used by this patient), offers the test, although they do say,

> *"This test was developed and its performance characteristics determined by ARUP Laboratories. The U.S. Food and Drug Administration has not approved or cleared this test; however, FDA clearance or approval is not currently required for clinical use."*

They also offer urine Lyme antigen and, as the Infectious Diseases Society of America guidelines say,

> *"Unvalidated test methods (such as urine antigen tests or blood microscopy for detection of Borrelia species) should not be used."*

Or billed for. My dad always said you judge a man by the company he keeps, but what about Lyme PCR on joints?

It seems to be a good test:

"In a U.S. study of 88 patients, B. burgdorferi sensu stricto DNA was detected in synovial fluid of 75 (85%) patients with Lyme arthritis (224). The PCR positivity rate was lower in patients who had received antibiotic therapy than in untreated patients. Of 73 patients who were untreated or treated with only short courses of oral antibiotics, 70 (96%) had a positive PCR in synovial fluid samples. In contrast, B. burgdorferi sensu stricto DNA was demonstrated in only 7 of 19 (37%) patients who received either parenteral antibiotics or oral antibiotics for more than 1 month."

But, to quote the IDSA:

"In a seropositive patient, a positive PCR test result on a synovial specimen adds increased diagnostic certainty. Positive PCR results for a joint specimen from a seronegative patient, however, should be regarded with skepticism."

So I told the orthopod to check the confirmatory tests for Lyme: these are ELISA (enzyme-linked immunosorbent assay) and Western Blot. When he had those, then he could send the patient to me. They were both positive. So Lyme it is. The patient has a reasonable exposure history, with travel to New Jersey and Europe, although never a tick bite or an erythema chronicum migrans rash. That's fine; an excellent exposure history is rare.

So now twenty-eight days of antibiotics.

Rationalization

Aguero-Rosenfeld, M.E., Wang, G., Schwartz, I. & Wormser, G.P. Diagnosis of lyme borreliosis. Clinical microbiology reviews 18, 484-509 (2005).

Captain, I Canna Get a Cure. I'm Givin' Her All I've Got.

I HAVE written about this case before. The patient is a renal transplant patient with red bump disease. For the past two months, he has experienced fevers, malaise, and progressive onset of red bumps on his skin. The history and physical examination revealed a history significant for IgA neuropathy and hypertension. His medications were cyclosporine and prednisolone; he had no history of smoking, alcohol use, illicit drug use, or exposure to HIV or sexually transmitted infections. His blood count and urinalysis were normal. Biopsy of the lesions showed necrotizing granulomas, with stains for acid-fast bacilli and fungi all negative.

The differential was puzzling. Syphilis was a reasonable guess, because it has multitudinous manifestations including cutaneous nodules and macules. But the patient's tests were both negative: we did a VDRL (Venereal Disease Research Laboratory) and FTA (fluorescent treponemal antibody) test.

The patient has cats, and in patients with AIDS, red bumps are usually a manifestation of bacillary angiomatosus, commonly known as cat scratch disease (CSD). However, the patient's CSD serology was negative and CSD can have stellate necrotizing granulomas. A review with the pathologist did not demonstrate stellate changes.

Cryptococcus can cause red bumps, often resembling *Molluscum contagiosum*. *Cryptococcus* is endemic in the Pacific Northwest, and when there are multiple skin lesions, the patients are systemically ill. The patient was not on calcineurin inhibitors, such as tacrolimus, which may by beneficial in cryptococcal disease. His serum cryptococcal antigen was negative.

Special stains were negative for lymphoma, which can manifest with necrotizing granulomas.

M. marinum grows in cool parts of the body because it cannot grow at 37 degrees. The lesions tend to be few in number and do not disseminate. Cultures were negative and the patient had no water exposures, the usual source for *M. marinum.*

Coccidiodomycosis can reactivate with immunosuppression and cause disseminated skin lesions, but the complement-fixation test was negative, as were fungal cultures.

So the first round of diagnostic testing was negative. Another biopsy specimen was obtained and sent to University of Washington, where molecular diagnosis using PCR with universal primers targeted at the 16S ribosome was performed, a truly amazing feat. It identified *Bartonella henselae*, the agent of CSD.

At first, the patient improved on ciprofloxacin and all of the red bumps and the fever went away. Unfortunately, after stopping the ciprofloxacin the fever and the red bumps returned and, during this time, his transplanted kidney failed to the point where his doctors elected to remove the it so that he would not have to stay on the transplant medications.

I figured that reversing his immunosuppression was what he really needed, so I treated the *Bartonella* again with ciprofloxacin, and again he got better, and a month or so after he had the nephrectomy, he stopped the antibiotics.

All was well. For a while. Then, you guessed it: the fever, cytopenia, and red bumps relapsed and the *Bartonella* PCR, which had reverted to negative, became positive again.

He had the five-star fever workup several months ago, including a transesophageal echo, and no focal source of the *Bartonella* was discovered. Why the relapse? Is there something funky about the patient's immune function? His antibody levels are normal, but his CD4 T-cell count is 140. In an HIV patient this would be the AIDS range, normal being greater than 500. But this patient is HIV-negative.

Now what? Don't take a temperature unless you want to treat a fever.

There are cases of chronic/relapsing *Bartonella* in patients with lymphocyte abnormalities outside of HIV: a couple of chronic lymphocytic leukemias are reported in the PubMeds.

What about the low CD4 count? Cause and effect is always an issue in ill people with low CD4s. Usually it is the stress of infection that leads to a low CD4, not the other way around.

His transplant antithymocyte globulin treatment was years ago, so you would expect his T cells to have recovered from that insult. CD4s do fall acutely, especially with the use of methyl-prednisolone.

> *"A general lymphopenia resulted in all patients with a mean decrease of 61 +/- 15% and an average nadir time occurring at 6 h. The decline from baseline was 76 +/- 17% for absolute number of CD4+ and 59 +/- 18% for CD8+ lymphocytes with an average nadir time at 6 h. Twelve patients exhibited a baseline CD4+ count to be less than 688 cells/mm3 (the low end of the reference range) and the lymphocyte count of all the patients fell below this value at the nadir. Six patients had a CD8+ lymphocyte count below 380 cells/mm3 (low end of the reference range) at baseline with 21 of the 23 patients exhibiting less than 380 cells/mm3 at the nadir time. At the time of nadir, the mean CD4+ and CD8+ counts were 156 +/- 105 cells/mm3 and 256 +/- 270 cells/mm3, respectively. In 17 of the 23 patients, the CD4+ count was below 200 cells/mm3 at the time of nadir."*

Transplant medications mess with the CD4, but the articles are way over my head:

> *"Overall, immunosuppression lowers CD4+CD25(high) FoxP3+/- Treg levels significantly in the periphery in renal transplant recipients. In addition, different immunosuppressive regimens have different impacts on CD4+CD25(high)FoxP3+ Tregs, a fact that may influence long-term allograft survival."*

Huh? Bit off more than I can chew there.

> *"CD4+CD25(high)FOXP3+ regulatory T cells (Tregs) play an important role in the maintenance of immunological self-tolerance*

by suppressing autoimmune responses and anti-tumor immune responses."

...I will forget even that before I am at the end of this sentence.

I can find nothing to suggest how long it will take for the T cells to return to normal--or if they will--once the medications are stopped.

So for now, I am going to put him back on the ciprofloxacin, look again for a focal source using transesophageal echocardiography, and follow the CD4s for the next few years.

Addendum

This time the patient was a cure.

Rationalization

Red Bumps (but no Lumps) in an Immunosuppressed Patient: Medscape article 741285.

Holmes, N.E., Opat, S., Kelman, A. & Korman, T.M. Refractory Bartonella quintana bacillary angiomatosis following chemotherapy for chronic lymphocytic leukaemia. Journal of medical microbiology 60, 142-146 (2011).

Maurin, M. & Raoult, D. Bartonella (Rochalimaea) quintana infections. Clinical microbiology reviews 9, 273-292 (1996).

Tornatore, K.M., Reed, K. & Venuto, R. 24-hour immunologic assessment of CD4+ and CD8+ lymphocytes in renal transplant recipients receiving chronic methylprednisolone. Clinical nephrology 44, 290-298 (1995).

Fourtounas, C. et al. Different immunosuppressive combinations on T-cell regulation in renal transplant recipients. American journal of nephrology 32, 1-9 (2010).

Sageshima, J. et al. Prolonged lymphocyte depletion by single-dose rabbit anti-thymocyte globulin and alemtuzumab in kidney transplantation. Transplant immunology 25, 104-111 (2011).

A Little Bit More for Santa Claus

I HAVE written on this topic a couple of times over the years, and it is a variation on the theme of the cellulitis that isn't.

The patient is a an elderly male, healthy, and living the good life of retirement. He develops cellulitis on his forehead—maybe a bite, maybe not—but he has a small patch of erythroderma. Over the next several days it worsens, so he seeks care and is started on cephalexin. But the erythroderma continues to spread, so he is admitted for progressive cellulitis. It progresses on IV antibiotics and so they call me.

He has no fevers, no chills, no sense of being ill, and his labs are normal.

His face is shiny beefy red, mostly on the right, but a bit now on the left. His lower lid is sagging. Kind of like two faces. I have seen this before, so after the usual history that reveals little, I ask pointed questions.

What are you putting on your face?

Aloe vera. As the infection spread, he would slather on more and more aloe vera and leave it on for longer periods of time. He was also using a product that was 99% aloe vera, so he was maximizing both the exposure time and antigen concentration.

Bingo. Contact dermatitis. And for what it is worth, it didn't look like cellulitis; it looked like—well, contact dermatitis. Don't ask me to describe the look with words; I don't speak dermatologic Latin. It is my impression that the lower lids of eyes sag with allergic dermatitis and swell up with infection to close the eye. Anyone know otherwise?

I have seen contact dermatitis treated as cellulitis with triple antibiotic ointment and with tea tree oil. The patients had some minor trauma they treated with topical agents, and you always apply ointments over an area a little bit larger than the lesion, so if you have an allergic reaction, it will grow as the area covered by the ointment expands.

The only thing I did not know is whether aloe vera has been reported to do this. Twenty seconds on the PubMeds, and the answer is yes. I will never, ever, miss the Index Medicus.

So I told them to do what I do best: nothing.

It was fun when I explained what I think is going on to the patient. He had this look of sudden understanding as he put all of the pieces together in his own mind, an "ah-ha" moment, which I have on occasion, but is always a hoot to see in others. Medicine, and ID, is satisfying in so many different ways.

Rationalization

Ferreira, M., Teixeira, M., Silva, E. & Selores, M. Allergic contact dermatitis to Aloe vera. Contact dermatitis 57, 278-279 (2007).

Persistently Continuous

FEVERS. I spend a lot of time thinking about fevers. Almost every patient I see can be coded 780.6.

The patient has multiple possible reasons for fever: coagulase-negative staphylococcal line infection, recent rupture of a ventricular wall myocardial infarction with emboli and patch repair, cholesterol emboli, wound infection, and prolonged hypotension with potential for gastrointestinal ischemia are all high on the list.

Despite identifying and potentially treating all of these conditions, the fevers persist.

The biggest worry is that the ventricular patch is infected, yet a repeat transesophageal echocardiogram is negative, the complete blood count is normal, and repeat blood cultures are negative. As best as can be determined without a re-op, the patch is not infected.

Fever curves are rarely helpful. How often the fever occurs and how high the fevers become are details that do not help make the diagnosis. Even malaria, the classic periodic fever, has rarely had the decency to recur with a classic frequency. The only infection where the fever pattern helps is relapsing fever--where the fever, well, relapses.

Except. There is one fever pattern that, if present, may be of help. The continuous fever. With most processes, the fever drops to normal at some point during the day, I suppose because the cause of the fever is intermittent.

As best I can tell, there is little on the topic of persistent/continuous fever, and the literature often uses it synonymously with FUO, fever of unknown origin. Most of the writings on the topic subsume it into the FUO genre. There is nothing but a few case reports on continuous fever, and it has the same differential diagnosis as FUO: cancer, infection (typhoid fever leading the list), and vasculitis. And—what I think this patient has—drug fever.

With drugs, the insult is continuous, and so is the fever. Or so I think, based on aggressive confirmation bias. While drug fever can be any pattern, continuous fever suggests drug fever. In the end, though, the two characteristics are that the patients are on a drug and that they have a fever.

> *"Unlike characterizations found in textbooks and review articles, we found relative bradycardia in a minority of cases reviewed; little risk associated with re challenge unless underlying cardiovascular disease was present; no characteristic fever pattern; a highly variable lag time between the initiation of the offending agent and the onset of fever; an infrequent association with either rash or eosinophilia; and no apparent association of drug fever with systemic lupus erythematosus, atopy, female sex, or advanced age."*

So I looked over the drug list, and the only potential culprit I noticed was the H_2 blocker for reducing stomach acid. For reasons that have more to do with a case I saw as a resident than with the robust medical literature, I associate H_2 blockers with drug fever. So I stopped it, and the fever was gone in about four half -lives.

True-True and unrelated or the real deal? Unless I rechallenge, which I am not about to do, I will never know. Someone who is not me needs to do a review of continuous fevers.

Rationalization

Hiraide, A., Yoshioka, T. & Ohshima, S. IgE-mediated drug fever due to histamine H2-receptor blockers. Drug safety 5, 455-457 (1990).

Mackowiak, P.A. & LeMaistre, C.F. Drug fever: a critical appraisal of conventional concepts. An analysis of 51 episodes in two Dallas hospitals and 97 episodes reported in the English literature. Annals of internal medicine 106, 728-733 (1987).

Ramboer, C. [Drug fever with cimetidine]. Lancet 1, 330-331 (1978).

Poll Results

The drug I most see that causes a fever is

- H_2 blocker. 9%
- sulfa. 15%
- vancomycin. 29%
- heroin. c18%
- pheromones. 24%
- Other answers 6%

  -ciprofloxacin.

Pessimistically Optimistic

The patient has MRSA bacteremia. Never a good thing, since MRSA is hard to kill under the best of circumstances, and with MRSA the circumstances are never the best.

S. aureus in the blood should warrant two questions: where did it come from, and where did it go (valves/bone)?

The answer to the second is apparently nowhere; the answer to the first is evidently an aortic graft. About a month ago she had a stent placed percutaneously into an aortic aneurysm. Now the CT scan shows air and fluid around the stent. The patient is, for a variety of reasons, not a surgical candidate.

For surgically placed grafts, surgical removal with extra-anatomical bypass gives the best outcomes, but there is a smattering of reports of salvage in patients deemed not to be surgical candidates—fewer for percutaneous stents.

The patient is not diabetic and only has smoking-related vascular diseases. "Only." Snort. She has defervesced and is clinically doing well, at least from the narrow perspective of infectious disease.

In my experience, I have cured a pair of aortic stent infections in my time, both due to MSSA (methicillin-sensitive *Staph aureus*). I treated them as if they had prosthetic valve endocarditis. Remember, though, that the most dangerous words in medicine are "in my experience," at least where therapeutic interventions are concerned. The downside here, too, is that we have MRSA, and most of the antibiotics for MRSA are crap compared to beta-lactams for MSSA.

But. I do have ceftaroline, a beta-lactam designed to kill MRSA. But. It has no track record for the infections you really want to use it for: endovascular infections and osteomyelitis. Who cares if it useful for community-acquired pneumonia and cellulitis? Damn near anything will work for those infections; I want to know about the difficult infections.

With this disease, I am going to have one shot to get it right. Go with the known crap that is vancomycin or the potentially superior but unproven beta-lactam? Especially with a vancomycin minimum inhibitory concentration (MIC) of 1.0, which isn't terribly high but isn't low either? Maybe it doesn't matter, since there is one study that suggests higher vancomycin MICs lead to higher failure rates with MSSA bacteremia treated with beta-lactams.

> *"However, even in patients with MSSA bacteremia treated with flucloxacillin, mortality was also higher if the vancomycin Etest MIC of their isolate was >1.5 μg/mL, compared with those with lower MIC isolates."*

I choose the beta-lactam and rifampin. Damned it I do, damned if I do.

Damn.

Rationalization

Gardner, G.P., Morris, M.E., Makamson, B. & Faizer, R.M. Infection of an aortic stent graft with suprarenal fixation. Annals of vascular surgery 24, 418 e411-416 (2010).

Setacci, C. et al. Management of abdominal endograft infection. The Journal of cardiovascular surgery 51, 33-41 (2010).

Calligaro, K.D., Veith, F.J., Yuan, J.G., Gargiulo, N.J. & Dougherty, M.J. Intra-abdominal aortic graft infection: complete or partial graft preservation in patients at very high risk. Journal of vascular surgery 38, 1199-1205 (2003).

Holmes, N.E. et al. Antibiotic choice may not explain poorer outcomes in patients with Staphylococcus aureus bacteremia and high vancomycin minimum inhibitory concentrations. The Journal of infectious diseases 204, 340-347 (2011).

The Worst Wedgie

THERE are times when they tell me to cure an infection that the literature suggests I can't. For whatever reason, the patient cannot have the definitive procedure--removal of the infection--and the doctors opt for medical therapy. And, on occasion, I have gotten lucky and cured the impossible: a yeast endocarditis or a graft infection. Even a blind pig gets an acorn once in awhile.

There is one infection where the medical cure rate is so close to zero that there is no point in even trying: pacer wire infections. I do not understand why the literature (and personal experience) is so grim for eradicating this infection without removing the wires. Prosthetic valve endocarditis, a moral equivalent, isn't that impossible to cure. For pacer wire infections, you always have to remove the system, which poses mechanical issues (pulling out a plug from the heart) and electrical issues (the patient doesn't have a cardiac rhythm without the pacer).

Once upon a time, when I was young and foolish, I had a patient with a *Serratia* pacer infection and the cardiologist said, "No way, no how, can I take out the pacer. We will kill him. You cure him."

So after he relapsed his first course of antibiotics, and the cardiologists demurred again, I thought—hey, the problem is the clot. Like central line infections, if I can strip off the clot with a little tissue plasminogen activator (tPA), perhaps I can get rid of the infection. Bad idea. The resultant huge surge of *Serratia* led to a day of sepsis in the intensive care unit. I explained beforehand the rationale, and apologized profusely afterward, and I must say the patient was very gracious. Then the cardiologist took out the pacer and the patient did just fine, thank you very much.

The current patient has a 2 x 3 cm vegetation on the pacer wire and multiple blood cultures for coagulase-negative *Staphylococcus*. "Gotta come out," I said, "and you probably do not have to worry about the clot." There is no ventricular septal defect or atrial septal defect seen on echocardiogram, and most of the time the lungs handle the clot breakup just fine. In twenty-one years, I have never had a pulmonary complication of pulling the pacer and the literature suggests,

> *"Vegetations on endocardial leads or right-sided cardiac structures ranging in size from 10 mm to 38 mm in their largest dimension were detected in 9 patients. All patients underwent successful transvenous removal of endocardial leads. Five of 9 patients (55%) had evidence of pulmonary embolism."*

So what is worse? Open procedure or potential emboli? We opted for the emboli, and I was not particularly worried as it was coagulase-negative *Staphylococcus*, which never causes pneumonia. Neither the Googles nor the PubMeds gives me a reference, although I did see one coagulase-negative *Staphylococcus* pneumonia years ago in the trauma ICU, in a patient with severe lung trauma.

And it embolized. Within forty-eight hours he was short of breath with a fever and a white cell count of 35,000 cells per microliter. A CT scan showed a clot wedged into, and partially obstructing, the pulmonary artery and a section of cavitating lung. I think the coagulase-negative *Staphylococcus* may have infected the infarcted lung, although I never had a culture to prove it. With lots of time and antibiotics, the patient slowly improved.

The first, and hopefully last, complication of pacer removal I have seen.

Rationalization

Meier-Ewert, H.K., Gray, M.E. & John, R.M. Endocardial pacemaker or defibrillator leads with infected vegetations: a single-center experience and consequences of transvenous extraction. American heart journal 146, 339-344 (2003).

Remembering Spam

I HAVE an affinity for the oddities that seem to abound in infectious diseases. I love the little factoids and risks, the peculiar presentations, that make each day a challenge. And above all, I love Google, because a nagging bit of memory can, in just a few seconds, become a fleshed-out fascinoma.

I am asked to see the patient for fevers. Fifty-eight, no chronic medical problems, he had been in the hospital with a fever and pneumonia a week ago, was discharged, and now comes in with a fever of 105 (really. 105). He also has increased shortness of breath, and is hypotensive but has no focal complaints.

Taking a history is limited by language difficulties, as he is from an island in Micronesia that has its own language. First, it's not pneumonia; it is heart failure on chest x-ray.

But why heart failure? The echocardiogram shows moderate mitral stenosis, and the cardiologist tells me that it looks rheumatic. Rheumatic heart disease is not a surprise, and the hypothesis is that he could not keep up with the demands of the high fever and went into congestive heart failure. Seems reasonable.

Then why the fever? The hospitalist tells me on the phone that the patient has cellulitis of the leg, and you betcha, the knee is cherry red and hot, just like a bit of Group A Streptococcus.So why the cellulitis?

Ah, there is the interesting part of the exam. On his lower

leg is a 4 x 5 inch verrucous red patch with some half-inch red bumps that are also verrucous. Red bump disease. How long has that been there, I ask?

Five or six years, slowly growing. I finish my usual exposure history: no pets, no other travel, no other exposures, although upon direct questioning he did note he was frequently in fresh and salty water back in Micronesia.

Red bump disease, chronic skin disease, Micronesia. Bing bing bing bing. The sound of a bell, not a search engine.

I read hundreds of articles a year for my Puscast and have almost 10,000 pdfs on my hard drive related to infectious diseases.

Every year I give a "best of the year" Grand Rounds covering the most interesting literature in the prior twelve months.

Even though I read and discussed these papers on my podcast, for at least two out of three I have no memory of reading them.

It is sad. But I remember the odd ones.

I had a memory of one or another atypical mycobacterium causing a chronic mycobacterial skin infections in Micronesia. The peculiarity in this case, what made it memorable, was the source of the non-tuberculous mycobacteria: water-filled shell holes from World War II that the natives would swim and fish in.

Sure enough, it was an acid-fast bacillus closely related to M. marinum:

> *"Spam disease" is a chronic, progressive skin disease of high prevalence on Satowan and is associated with taro farming and contact with World War II-era bomb craters."*

You can see pictures that are identical to my patient in the paper cited in the references.

Why did I remember this, of all the things I read last year? Got me.

I still wonder where these memories come from.

The patient never would follow-up, so I have no idea what became of him.

Rationalization

Lillis, J.V. et al. Sequelae of World War II: an outbreak of chronic cutaneous nontuberculous mycobacterial infection among Satowanese islanders. Clinical infectious diseases : an official publication of the Infectious Diseases Society of America 48, 1541-1546 (2009).

Mutiplying Entities

THE patient presents with low-grade fevers and a half dozen dewdrops on a red base on the arms and chest. Having had an autologous bone marrow transplant for lymphoma a decade ago made her at risk for VZV (*Varicella zoster* virus, aka chicken pox). Her primary care doctor asked to me to look at the rash, and it sure looked like chicken pox to me. The patient said she had chicken pox as a child.

VZV, under these circumstances, is usually felt to be reactivation from the primary illness as a child, but in talking to the spouse, I discovered he had a case of shingles. So is this a new infection or a reactivation? Can't say, but I am betting on a new infection.

Reinfection can happen—in patients, both normal and immunoincompetent, and in doctors as well:

> *"A 32-year-old physician with a history of chickenpox at age 5 and seropositivity to varicella-zoster virus (VZV) at age 30 developed fever and vesicular rash 14 days after examining an immunocompetent patient with localized herpes zoster ophthalmicus. Vesicular viral culture grew VZV, and the physician was diagnosed with VZV reinfection."*

Often the disease caused by reinfection is less severe than an initial infection, or even asymptomatic with the only sign being an increase in the antibody titers. Of interest, this patient's VZV serology is negative.

The worry with VZV is pneumonia, and the patient has had three weeks of nonproductive cough. A chest x-ray shows perhaps a few largish peripheral nodular infiltrates, not the pattern of VZV pneumonia. So a CT scan is done and confirms multiple peripheral nodules. We find out that the symptoms started shortly after the patient returned from Arizona. Almost certainly, *Coccidioidomycosis* (aka Valley Fever)' bronchoscopy results are pending.

I am an Occam's kind of guy; I always say *entia non sunt multiplicanda praeter necessitatem.* Yeah, I speak Latin on rounds. *E pluribus unum*, dude. But I think that the patient has two processes: a VZV reinfection and *Coccidioidomycosis* pneumonia. I don't think I am multiplying entities beyond necessity.

And I wasn't.

It was *Coccidioidomycosis* as well, and both were treated with success.

Rationalization

Johnson, J.A., Bloch, K.C. & Dang, B.N. Varicella reinfection in a seropositive physician following occupational exposure to localized zoster. Clinical infectious diseases : an official publication of the Infectious Diseases Society of America 52, 907-909 (2011).

Junker, K., Avnstorp, C., Nielsen, C.M. & Hansen, N.E. Reinfection with varicella-zoster virus in immunocompromised patients. Current problems in dermatology 18, 152-157 (1989).

A Return to the Bad Old Days

To my way of thinking, the old days were the bad days. There is no shortage of problems in the world, but for the most part, it has been one of slow improvement over time. I remember from a textbook that there are some scrolls from around 5000 BC where the writer complains about the (then) current times, how the younger generation has no respect, and how things were better back in the day. I try to keep that in mind when I feel an urge to grouse about kids today.

Nope. I can think of nothing that is worse now when compared to the past. Except for global warming. And Medicare reimbursement. But everything else in my life, professional and personal, is better now than it was before.

When I started my practice in the last decade of the last century, the hospital was filled with young men dying of the various complications of the last stages of HIV. I do not see that many HIV opportunistic infections anymore, which is a good thing. This week, back from the Infectious Diseases Society of America conference, was a return cross-country and back in time.

There are three patients in the hospitals where I work with complications of HIV. They all share a common feature: no insurance. USA USA USA. We are number thirty-seven, just ahead of Slovenia. Go, USA. I do not pretend to have even a hint of a solution to the mess that is U.S. health care, since my hubris does not extend beyond ID, but I grow ever so weary of people sickening and occasionally dying, and going bankrupt in the process, from lack of access to health care.

I suppose I should look on the bright side: it is good for my professional experience to see severe, rare, yet mostly preventable illnesses, but there is no satisfaction to be gained in pyrrhic victories.

One of the patients has severe *Pneumocystis jiroveci* pneumonia (PJP) and I expect that she will get better with time, despite the initial delay in providing her with the best therapy. I guess another thing that is not as good as it was back in the day, is drug shortages. The last decade has seen an increased short-term and long-term lack of availability of many drugs, including antibiotics. We have had no IV trimethoprim/ sulfamethoxazole for about a year now, which is the preferred drug for the treatment of PJP. Just in the last month or so we have been able to get it on special order, but it takes a day, and so we had to use pentamidine instead. A less effective and more toxic alternative. Great.

But it did provide an opportunity for us to use a test for the first time for PJP. On admission, the chest x-ray showed heavy involvement. Usually those who need intubation have lactate dehydrogenase (LDH) values in the 600s. PJP lives on type II pneumocytes, and type II pneumocytes make LDH. As the white cells kill the PJP, they also kill the adjacent pneumocytes, releasing LDH. Of course, that requires the ability to mount an inflammatory response, which is not always present in HIV. So when the LDH came back only 280—and it is not the greatest test—we had to ask ourselves, well, what else could be going on?

"In HIV-negative patients the sensitivity of LDH elevation was 63% and specificity 43%. In HIV-positive patients sensitivity was 100% and specificity 47%. The overall accuracy of LDH for the diagnosis of PCP was 52%, 51% in HIV-negative and 58% in HIV-positive patients."

Is it H1N1 or some other equally bad atypical/viral pneumonia?

So the team sent off a beta-D-glucan test, which is diagnostic for fungal infections, and it came back off the charts. When I was a resident, we thought PJP was a protozoan, but instead it is a yeast-like fungus. It's a damn peculiar fungus at that, being treated with trimethoprim/ sulfamethoxazole instead of an azole or amphotericin, although caspofungin does have its uses.

Also, the CD4 cell count was 5 cells per microliter (normal greater than 500). So I attribute the unusually low LDH relative to the severity of the pneumonia to the lack of an immune response. No functioning white cells to kill the PJP or the pneumocytes.

When the patient required intubation, the specimen came back positive for PJP, rounding out the diagnostic testing. Once the trimethoprim/ sulfamethoxazole was airlifted in, she slowly improved. I expect a good outcome, although I remember the bad old days when a pneumothorax was not unexpected.

Rationalization

Armstrong-James, D. et al. A trial of caspofungin salvage treatment in PCP pneumonia. Thorax 66, 537-538 (2011).

Matsumura, Y. et al. Quantitative real-time PCR and the (1-->3)-beta-D-glucan assay for differentiation between Pneumocystis jirovecii pneumonia and colonization. Clinical microbiology and infection : the official publication of the European Society of Clinical Microbiology and Infectious Diseases 18, 591-597 (2012).

Vogel, M. et al. Accuracy of serum LDH elevation for the diagnosis of Pneumocystis jiroveci pneumonia. Swiss medical weekly 141, w13184 (2011).

BAM

So not every case has a denouement, at least not at the moment. I try to remember it is the journey, not the diagnosis, although I have to admit that there is nothing in life as satisfying as making a diagnosis. Everyone derives pleasure from different aspects of their job, and for me, it is figuring something out. The problem with medicine is that all too often, for a variety of reasons, I do not get a diagnosis. As long as the patient improves, I can live with that.

This week I saw an elderly male with fevers and a rash with no diagnosis after twenty-four hours, but he looked ill and his inflammatory parameters were elevated. His white blood cell count and procalcitonin were up, but the rest of his admit studies were otherwise negative for a focal infection.

I go through the usual routine, reading the chart, looking at the chest x-ray (slight patchy infiltrates not present as an outpatient), and so forth, and then go into the room to see the patient.

BAM. On occasion, when seeing a patient for the first time, there is some finding on history or, in this case, the physical, that grabs your attention and is the hinge-point for the differential diagnosis. I do not know about you, but it takes a lot of willpower to hold back and to go through the usual routine and not to jump straight to the elephant in the room. I try to consciously not fall prey to premature closure, a common reason for mistakes.

> *"The most common cognitive problems involved faulty synthesis. Premature closure, i.e., the failure to continue considering reasonable alternatives after an initial diagnosis was reached, was the single most common cause. Other common causes included faulty context generation, misjudging the salience of findings, faulty perception, and errors arising from the use of heuristics."*

It is not easy to do.

What grabbed my attention? Both of his eyes were bloodshot red, and outside of a college dorm, there are few things that give this kind of conjunctival suffusion. He also has a macular rash, dime-size, all over his body.

So I take the usual history, taking my time. I don't want to. I want to ask leading questions about the eyes and get my answer.

I study the history of the present illness, the past medical history, the review of systems, and the allergies: habits slowly go by. Then the fun.

Exposures? He is around several grandchildren who are healthy.

So that makes adenovirus a possibility, although that should not raise the procalcitonin and white count.

And he has no cough to account for the x-ray. He is too old for *Mycoplasma*, which could do this as well.

Animals? Just dogs and cats from friends and family. No odd animals.

Travel? It started in the Bay Area. *Coccidioidomycosis*? *Coccidioidomycosis* causes *erythema nodosum*, which this isn't, and this is not a coughing illness. Not thrilled with that idea, and it does not account for the eyes.

Water? Lives on a boat on the Columbia, was on the Bay in San Francisco, and while he likes to swim and scuba in the rivers of Portland, he has not been in the water recently. Crap. I thought for sure with those eyes it would be Leptospirosis, but the water exposure is not what I would like. Close. And if it is Leptospirosis, it is odd with the rash and with the lack of hepatic and renal involvement. But but but, Leptospirosis is a biphasic illness and the extensive organ involvement usually occurs in the second phase of the illness. So it could be early Leptospirosis, and patients do not like to read the textbooks when they decide to develop their symptoms.

So as of now I am saying Leptospirosis > adenovirus > Mycoplasma.

Everything is pending, so we shall see.

Postscript

Never got a diagnosis, all studies were negative, and he got better. Such is life.

Rationalization

Leptospirosis: http://emedicine.medscape.com/article/220563-overview

Graber, M. L., Franklin, N. & Gordon, R. Diagnostic error in internal medicine. Archives of internal medicine 165, 1493-1499 (2005).

Poll Results

I gain my satisfaction in medicine by

- making a diagnosis. 22%
- helping the patient to get better. 33%
- the interactions I have with my patients and with their family. 4%
- going home. These days medicine sucks. 9%
- taking medicines. Mmmmmmm. Zoloft. 29%
- Other Answers 2%

confirming that an adverse effect is caused or exacerbated by low nitric oxide.

Twelfth Night. Sort of.

Be not afraid of the diagnosis: Some are born to diagnose, some achieve a diagnosis, and others have a diagnosis thrust upon them. —Willy Shakespeare

THE last of the three options is the least fun, but it is interesting to see the previously diagnosed patients who occasionally come my way. It is a, "Yep, you sure have X" consult, but not always. Sometimes I agree with the diagnosis; sometimes I do not. Microbiology is unreliable, and sometimes the odd diagnosis is a contamination and not a pathogen. One of the guiding rules I had as a resident was, "The emergency room diagnosis is always wrong." As a consultant, I assume that everyone is always wrong, including me. It is a good, if somewhat time-consuming, approach if you do not want to get lulled into a false diagnosis. *Doveryai, no proveryai* dude.

The patient presents with six months of acne. Not unusual if seventeen, but not so typical in one's seventies.

The lesions were only on the back of the neck and the cheeks. They were biopsied; the pathology demonstrated granulomas and the cultures grew MAI—*Mycobacterium avium*. The patient was sent to me, the lesions all gone on a course of azithromycin.

I took my usual history, reviewed the data, and looked at the normal skin. Yep. You sure had MAI folliculitis.

But.

How often does MAI cause folliculitis in immunologically normal people?

Almost never. As with all organisms, there is a smattering of cases mostly in the immunoincompetent, but in the spectrum of acid-fast bacilli that cause red bump diseases, MAI is not high on the list.

> *"Twenty-eight biopsy specimens corresponding to 27 patients with cutaneous infections due to NTM were reviewed. Eighteen biopsies corresponded to normal hosts (14 Mycobacterium marinum, 2 Mycobacterium chelonae, 1 Mycobacterium terrae and 1 Mycobacterium gordonae) and 10 biopsy specimens were obtained from 9 immunosuppressed patients (3 Mycobacterium chelonae, one of which had two biopsies, 1 Mycobacterium abscessus, 2 Mycobacterium kansasii, 1 Mycobacterium marinum, 1 Mycobacterium avium complex and 1 Mycobacterium simiae)."*

And then why? When I think MAI, I think water, and MAI is found in tap water and is a cause of hot tub hypersensitivity pneumonia, but not hot tub folliculitis. To make matters more annoying, the patient has not been in a hot tub in years; the only water or exposure to the head and neck was the shower. Hardly an impressive risk factor.

So the histopathology, culture, and response to therapy support a diagnosis with no good risk factors for a disease that is rarely reported in normal hosts, and he is better as a result of the antibiotics, so I have no reason clinically to suspect another diagnosis.

It is not the way I want a diagnosis thrust upon me. Not as clean as I would like, but that is medicine, and what is a poor ID doc to do? If this were played upon a stage now, could I condemn it as an improbable diagnosis?

Rationalization

Bartralot, R. et al. Cutaneous infections due to nontuberculous mycobacteria: histopathological review of 28 cases. Comparative study between lesions observed in immunosuppressed patients and normal hosts. Journal of cutaneous pathology 27, 124-129 (2000).

Meningitis and Lips

It has been a very slow week. So far, two consults in the last four days. Nothing much to write about; that suits me fine, as I have been ill since last Friday, and given the incubation period, I probably picked it up in Boston. No, not a sexually transmitted disease; get your mind out the gutter. Some upper respiratory infection that is just bad enough to turn my brain into pudding. Good thing work was slow, since I did not feel like passing whatever it is on to my patients. Today was the first day in a week I have felt close to normal and I actually had a consult.

An elderly diabetic male presents with fevers and with an altered mental status. A spinal tap shows elevated protein, low glucose, and increased white blood cells: this suggests bacterial meningitis. But the gram stain had gram-negative rods. Not a typical gram stain, although the elderly are the one group who can get a gram negative meningitis, though they usually have a focus of infection elsewhere. In this case, the blood and spinal fluid grew *E. coli*, but the workup for a source yields nothing.

A spontaneous case of *E. coli* bacteremia with meningitis. Not satisfying. All I can do is give a course of antibiotics and watch the patient do unexpectedly well. Then the patient developed a cold sore. *Herpes simplex* virus. Could this be HSV encephalitis as well? No.

HSV will commonly reactivate with bacterial meningitis, although I will be darned if I can find the reference. The one I could find is not available through our library. At the end of the last century I had a patient with pneumococcal meningitis who developed a cold sore, and with the help of Index Medicus found the references to indicate that central nervous system inflammation often results in HSV reactivation. You would think that it would be a piece of cake to recapitulate the search with Google and PubMed; wouldn't you? My Google-fu must be inhibited by the low levels of interferon from my infection.

HSV has as long been referred to as fever blisters, and more than one cause of fever can cause HSV to blossom:

> *"Artificial fever induced by physical means (153, 154) or by injection of bacterial products (155, 156) is complicated by herpes labialis in 40 to 70 percent of cases.*
>
> *Among diseases in which fever blisters are a frequent complication are malaria, relapsing fever, viral influenza, typhus, and other rickettsioses, and infections produced by pneumococci, meningococci, and streptococci (157-160). However, herpes labialis is so unusual in the course of tuberculosis, smallpox, primary atypical pneumonia, brucellosis, and typhoid, that its presence argues strongly against the clinical diagnosis (157-160) of these diseases. The rarity of fever blisters in typhoid contrasts with their frequency in patients given typhoid vaccine and raises an interesting question about the role of endotoxemia in the pathogenesis of typhoid.*
>
> *There is relatively little interest in such clinical trivia as fever blisters in this antibiotic age. Their presence or absence is not often recorded in clinical charts and retrospective reviews of hospital records will not often yield reliable information."*

The grumpy complaint from 1960, about the lack of interest by youngsters in clinical trivia, marks the writer as a kindred spirit.

Kids today. They just don't have what it takes.

Rationalization

Laguna-del Estal, P. et al. [Bacterial meningitis due to gram-negative bacilli in adults]. Revista de neurologia 50, 458-462 (2010).

Weerkamp, N., van de Beek, D., de Gans, J. & Koehler, P.J. Herpes reactivation in patients with bacterial meningitis. The Journal of infection 57, 493-494 (2008).

Bennett, I.L., Jr. & Nicastri, A. Fever as a mechanism of resistance. Bacteriological reviews 24, 16-34 (1960).

Dehydrated Pneumonia

THE patient is elderly, demented, and with no ability to care for himself. He has a fever and low blood pressure. He has the usual workup in the ER for a cause of the fever and nothing is found, and he is admitted to the hospital for fever and dehydration (his serum sodium is 158 milliequivalents per liter; normal is less than 145). Every time I say dehydration, I hear my renal attending from thirty years ago saying, "Camping food is dehydrated. Humans are volume short." Odd, since REI sells their Volume Short line of backpacking food.

His fevers persist, with a leukocytosis, and after twenty-four hours they call me. While I am seeing the patient another chest x-ray is obtained, and when I am done with my history and physical, there is now a right lower lobe infiltrate. Not cause and effect, by the way. So he has pneumonia, a common cause of symptoms like this.

So the chest x-ray was negative on admission because the patient was volume short, and it delayed the findings. Right? The infiltrate blossomed like a pneumonic flower. So I was always taught, and it doesn't make much sense to me. I have trouble believing that being volume short will delay the edema and white blood cells that show up on a chest x-ray. Probably it is a combination of time, an insensitive study, and a whopping confirmation bias.

The chest x-ray is a lousy study for the diagnosis of pneumonia:

> *In patients admitted with a clinical diagnosis of CAP [community acquired pneumonia], the initial chest radiograph lacks sensitivity and may not demonstrate parenchymal opacifications in 21% of patients. Moreover, greater than half of the patients admitted with a negative chest radiograph will develop radiographic infiltrates within 48 hours.*

And if you do a CAT scan on patients with suspected community-acquired pneumonia, you will find additional cases missed by the chest x-ray:

> *Simultaneously obtained chest radiographs were compared with HRCT scans for 47 patients with clinical symptoms and signs suspicious for CAP, HRCT identified all 18 CAP cases (38.3%) apparent on radiographs as well as eight additional cases (i.e., 55.3%).*

In dogs,

> *We conclude that in a canine model, moderate dehydration has no effect on the radiologic or histologic features of Streptococcus pneumoniae pneumonia.*

So is there any human data? There is one unconvincing retrospective study that suggested,

> *Elevated admission BUN level and higher fluid volume administered in the first 48 hours of admission were associated with worsening radiographic findings of pneumonia after hydration."*

It would be an easy calculation for someone a lot smarter than me: At what osmolality will there be a decrease in the fluid flowing into the alveoli during pneumonia? I bet it will be when the blood has the thickness of pudding.

I suspect that dehydration delaying pneumonia findings on an x-ray is a myth, like the one that says we use 10% of our brain (the presidential debates notwithstanding). Or that Eskimos have 300 words for snow— although in Oregon, we have 300 words for rain, and they all start with damn.

Rationalization

Hagaman, J.T., Rouan, G.W., Shipley, R.T. & Panos, R.J. Admission chest radiograph lacks sensitivity in the diagnosis of community-acquired pneumonia. The American journal of the medical sciences 337, 236-240 (2009).

Syrjala, H., Broas, M., Suramo, I., Ojala, A. & Lahde, S. High-resolution computed tomography for the diagnosis of community-acquired pneumonia. Clinical infectious diseases : an official publication of the Infectious Diseases Society of America 27, 358-363 (1998).

Hash, R.B., Stephens, J.L., Laurens, M.B. & Vogel, R.L. The relationship between volume status, hydration, and radiographic findings in the diagnosis of community-acquired pneumonia. The Journal of family practice 49, 833-837 (2000).

Caldwell, A., Glauser, F.L., Smith, W.R., Hoshiko, M. & Morton, M.E. The effects of dehydration on the radiologic and pathologic appearance of experimental canine segmental pneumonia. The American review of respiratory disease 112, 651-656 (1975).

Didn't See that Coming. Again.

The nice thing about being a doctor is the ample opportunity to be blindsided. Even when you might think you know what is going on, and especially when you don't, the real answer will come out of left field. You—or in this case me—can do little beyond shaking my head and mumbling, "Didn't see that coming."

The patient is a middle-aged diabetic who is admitted with fevers, is hypotensive, and the source is maybe a diabetic foot infection. I was called when the blood cultures grew *Candida glabrata*, a yeast, in one of four bottles, drawn after the central line was placed. He is already better with fluids and antibiotics.

The patient looked fine clinically. The foot did not look infected, and the review of systems and a physical exam, as well as the rest of the cultures, were negative for a source of the *Candida*. He was uninsured, his kidney function was marginal—elevated creatinine at 2.2 mg/dL—and his symptoms were totally nonfocal. For all of these reasons, I did not order a CT scan due to the risk for kidney toxicity. I thought maybe the yeast was a contaminant, maybe from the line, but I was uncertain. Still, hard to ignore yeast in the blood of a diabetic, so he received a course of higher-dose fluconazole, an antifungal.

Fast-forward six weeks. The patient is admitted with a scrotal and penile erythroderma and swelling. I am called because this time the urine is growing *C. glabrata.* Again, a review of systems is negative, and the skin looks like diaper rash. His creatinine has improved, so we CT abdomen and pelvis. I'm thinking maybe *Candida* pyelonephritis, a disease that can be clinically silent, although difficult to prove, and perhaps he seeded his kidneys with the fungemia.

And the CT shows?

Multiple small pelvic abscesses, which are incised and drained. Cultures? Only *C. glabrata.* No trauma, no symptoms, no reason for *Candida* pelvic abscesses. No risk factors at all. As best as can be determined, this is de novo.

There is the smattering of similar cases in the literature, but there was no reason for the abscess and/or fungemia: surgery or one of the more typical risk factors for a *Candida* infection. Spontaneous *Candida* abscesses are unreported.

Didn't see that coming.

Again.

Rationalization

Malani, A. et al. Candida glabrata fungemia: experience in a tertiary care center. Clinical infectious diseases : an official publication of the Infectious Diseases Society of America 41, 975-981 (2005).

Wiesenfeld, H.C., Berg, S.R. & Sweet, R.L. Torulopsis glabrata pelvic abscess and fungemia. Obstetrics and gynecology 83, 887-889 (1994).

A Great Diagnosis (at Least In My Fevered Imagination)

I MADE a great diagnosis today. Of course, it will never be confirmed, as so many of my great diagnoses are. But why worry about reality proving me wrong? As the actor Adam Savage said, or at least his T-shirt, "I deny your reality and substitute my own." Works for me.

The patient is an elderly male with no past medical history, who is admitted with three days of fevers, rigors, and a papular rash sparing the palms and soles.

His infectious disease exposure history is significant for an adopted, stray, mostly outdoor cat that occasionally scratches him, and the illness started while on a long weekend to the Oregon coast.

He has a low white blood cell count: 2.1 with 23% bands, a low hemoglobin at 10 mg/dL, and a very low platelet count of 32,000 (normal is 150,000-450,000). His liver tests are twice the level of normal, the CT scan of his abdomen is negative, and the chest x-ray shows no infiltrates.

This is what is called a "doxycycline deficiency state": multi-organ involvement, febrile rash with no focal pus, cytopenias, and elevated liver function tests. The bugs that do this are *Rickettsia*, *Ehrlichia*, *Coxiella*, and *Spirochetes* (syphilis and leptospirosis), all of which are treated with doxycycline.

Which one is this? Judging by his relative lack of history, the only exposure would be for *Rickettsia*, and with the quasi-feral cat, I bet *R. felis*. It could be the mystery *Rickettsia 364D* reported in the CID this year, or even a new *Rickettsia*. There are so many, and it seems like there is a new one reported every month infecting some poor sap. No other pathogen seems likely by clinical presentation, and the great Pacific Northwest is free of *Ehrlichia*.

R. felis, found in cats and in opossums, causes a Rocky Mountain Spotted Fever- like illness, which he has. Mystery *Rickettsia* 364D causes a Rocky Mountain Spotted Fever-like illness, but with an eschar, which this patient does not have. All Rickettsia infections are reportable, and the State tells me they have had no reported cases of murine typhus (*Rickettsia typhi*) for a long time, probably because no one is looking.

Of course, acute serologies are sent to the State, and will be back in three weeks, and will be almost certainly negative. Convalescent serologies will, despite my best efforts, never be sent, and I will probably never know the real diagnosis. The patient is now on doxycycline, and we shall see how he does.

That's my story, and I am sticking to it.

Postscript

The patient improved on antibiotics or because of antibiotics, I do not know, as initial serologies were negative and a repeat was never done.

Rationalization

Perez-Osorio, C. E., Zavala-Velazquez, J. E., Arias Leon, J. J. & Zavala-Castro, J. E. Rickettsia felis as emergent global threat for humans. Emerging infectious diseases 14, 1019-1023 (2008).

Shapiro, M.R. et al. Rickettsia 364D: a newly recognized cause of eschar-associated illness in California. Clinical infectious diseases : an official publication of the Infectious Diseases Society of America 50, 541-548 (2010).

Two Down, So One to Go

Hospitalists are good and bad. Good, in that my hospitals have bright doctors who provide excellent care. Bad, in that my hospitals have bright doctors who provide excellent care. Much of the time they do not need an infectious disease consult and they have a nasty habit of learning from prior consults, so that the next time a relatively common problem gets admitted, they remember what to do. Work was busier back in the day when inpatient care was provided by the outpatient docs, who had Fear Uncertainty Doubt (FUD), whether they admitted it or not, and FUD leads to consults.

FUD with style is what makes a good consultant.

As a result of smart hospitalists, much of my practice has become the unusual bug or the un-attributable fever. I'm not complaining, mind you. The work volume may be less, but the cases are always interesting and challenging.

This week I had a pair of very common bugs in an extremely unusual place. Both patients had immunologic problems: one had recurrent infections due to combined variable immunodeficiency (CVID) that wasn't being treated with intravenous immunoglobin with any regularity. The other had HIV with a low CD4 cell count.

Both had destructive, cavitary pneumonias, the HIV patient with an empyema (pus in the chest cavity) as well. The bug? Group A *Streptococcus*, *S. pyogenes*. I have not seen a Group A streptococcal pneumonia this century, and maybe they are only two I have seen, ever. Maybe. My memory is not what it once was, filled mostly with the lyrics of old pop songs, but as I write I cannot recall a prior case. If you are new in practice, start a database of every case you see. You will not regret it when you are older and sliding into senescence.

Group A *Streptococcus* does cause the occasional pneumonia, and the patients are usually very ill, with abscess formation the rule. Both patients also took NSAIDs for the fever and pain. NSAIDs plus pneumonia equals badness:

> *"However, they more often developed pleuropulmonary complications, such as pleural empyema and lung cavitation (37.5% vs 7%; P = .0009), and had a trend to more-invasive disease, with a higher frequency of pleural empyema (25% vs 5%, P = .014) and bacteremia, especially in those not having received concomitant antibiotics (69% vs 27%, P = .009). Nevertheless, the patients in the NSAID group had no more severe systemic inflammation*

or remote organ dysfunction. In multivariable analyses, NSAID exposure was independently associated with the occurrence of pleuropulmonary complications (OR, 8.1; 95% CI, 2.3-28).
CONCLUSIONS:
Our findings suggest that NSAID exposure at the early stage of CAP is associated with a more complicated course but a blunted systemic response, which may be associated with a delayed diagnosis and a protracted course."

Is it the NSAIDs, the host, or the organism? Yes. All three. There is never one reason why people get infections, just as there is never one intervention that prevents infection. The downside is that bad things come in threes, so sometime in the next twenty-five years I am going to see a third case of Group A *Streptococcus* pneumonia.

Guaranteed.

Rationalization

Pan-Hammarstrom, Q. & Hammarstrom, L. Antibody deficiency diseases. European journal of immunology 38, 327-333 (2008).

Frieden, T.R., Biebuyck, J. & Hierholzer, W.J., Jr. Lung abscess with group A beta-hemolytic Streptococcus. Case report and review. Archives of internal medicine 151, 1655-1657 (1991).

Voiriot, G., Dury, S., Parrot, A., Mayaud, C. & Fartoukh, M. Nonsteroidal antiinflammatory drugs may affect the presentation and course of community-acquired pneumonia. Chest 139, 387-394 (2011).

Post Hoc Ergo Propter Hoc

SEVERAL years ago, the American Heart Association simplified its recommendations for the use of prophylactic antibiotics to prevent endocarditis. There used to be a long list of heart conditions that would automatically lead to prescribing antibiotics before any dental procedure. But in 2017 many diseases were thrown off the list, no longer requiring prophylaxis.

The data to suggest benefits from prophylactic antibiotics was mostly animal models combined with paranoia. After the initial guidelines were published, there were epidemiologic studies that all suggested that prophylactic antibiotics did not prevent endocarditis, and over the years, the indications were narrowed.

Not the least bit unreasonable, since low-grade bacteremia is the norm. Brush, floss, have a bowel movement, squeeze a zit, and you will have a touch of bacteremia. Given the rarity of endocarditis, the ubiquity of bacteremia, and many people going to the dentist every six months, it would be very difficult to prove that a given dental procedure resulted in a valve infection and not from the bacteremia of life. Giving antibiotics is not without harm, and the odds of dying from penicillin anaphylaxis are greater than dying from dental- induced endocarditis. But who can deal with odds?

Today's patient has mitral valve prolapse and for years has been taking antibiotics for endocarditis prevention prior to dental cleaning. With the new guidelines, the most recent visit was not covered with amoxicillin and wouldn't you know? The patient developed endocarditis in the month after the dental cleaning.

Was it due to the dental cleaning? No way to know. Odds are against it. *S. viridans* endocarditis is a disease of the elderly, and the patient probably developed endocarditis from life. Not that I could convince him otherwise.

What is more convincing: a single anecdote or several epidemiologic studies that demonstrate the lack of association between dental work and endocarditis? The anecdote. Always.

Stories, especially medical stories of success or a complication of a therapy, are powerful for both the patient and for the physician, and it takes a great deal of effort to remember that the three most dangerous words in medicine are, "In my experience," at least as far as therapeutic interventions are concerned. We are creatures of the post hoc ergo propter hoc, reasoning imperfectly at baseline and rarely recognizing it. So I always try to go with the literature in deciding on interventions. In my experience it is more reliable.

Rationalization

Wilson, W. et al. Prevention of infective endocarditis: guidelines from the American Heart Association: a guideline from the American Heart Association Rheumatic Fever, Endocarditis, and Kawasaki Disease Committee, Council on Cardiovascular Disease in the Young, and the Council on Clinical Cardiology, Council on Cardiovascular Surgery and Anesthesia, and the Quality of Care and Outcomes Research Interdisciplinary Working Group. Circulation 116, 1736-1754 (2007).

Poll Results

My favorite logical fallacy is

- post hoc ergo propter hoc. 21%
- epluibus unum. 9%
- argumentum ad ignorantiam. 42%
- cum hoc ergo propter hoc. 5%
- confundo. 16%
- Other Answers 7%
    -cogito ergo sum.

Postscript

This was originally written in 2011. I compiled the book in 2019, and now there are studies to suggest that relaxing prophylaxis has increased the incidence of endocarditis for some cardiac anomalies. More to come, I am sure.

Relapse or Reinfection

THREE years ago the patient had a liver abscess for no good reason that I can tell. He was treated at "Outside Hospital" with a course of intravenous and oral antibiotics after drainage. The records from Outside Hospital say that the infecting organism was a *Streptococcus,* not Group D. Many labs do not speciate *Streptococci*, as it costs too much money to get a full name, and the complete name often doesn't matter. Well, it does to me. I like the full name of the bug, and sometimes the name of the bug gives a hint as to where it came from. That it is not Group D means it wasn't a *bovis*, so I—or rather he—doesn't have to worry about colon cancer.

Fast-forward three years. He was doing fine and was heading toward an elective hernia repair when the pre-op blood work showed an increased white blood cell count. One thing led to another, as one thing often does, and they did a CT scan, and there—right there—was another liver abscess.

He denied any symptoms initially, but upon closer questioning, admitted he had had several weeks of night sweats and fatigue, but no fevers or pain over the liver.

A tap of the abscess grew *S. anginosus*. No surprise; this is a bowel *Streptococcus* that is often found in liver abscesses. Again, no good reason for the abscess in the liver. No biliary problems, no bowel problems, no dental problems, no reason for a bacteremia, or for a local infection.

Reinfection? Or Relapse?

I cannot believe he had an occult liver abscess for three years without any symptoms, and there are a mere two cases in the PubMeds of recurrent *Streptococcal* liver abscesses, and a few others due to *K. pneumoniae.*

It is a curiosity. The patient wanted to know why he was experiencing the recurrent abscess, and I was left with a Gallic shrug and with my usual reason: bad luck. At least the patient didn't respond with the usual: "It figures; I get all the bad luck. If something bad happens, it always happens to me." They never say, "How odd, I am always lucky; most peculiar this happened to me."

And me? I am always lucky. I get to do ID for a living.

Rationalization

Chua, D., Reinhart, H H. & Sobel, J.D. Liver abscess caused by Streptococcus milleri. Reviews of infectious diseases 11, 197-202 (1989).

Tomiyama, R. et al. [Clinical features in six patients with liver abscess caused by Streptococcus milleri]. Nihon Shokakibyo Gakkai zasshi = The Japanese journal of gastro-enterology 98, 1060-1064 (2001).

Poll Results

I have

- good luck. 11%
- bad luck. 3%
- equal amounts of both. 11%
- no such thing as luck. 11%
- the great luck to read this book. 66%

There Is Much to Be Thankful For

THANKSGIVING is a good time for ID.

Turkey: salmonella, *Campylobacter*, Psittacosis, avian flu, *Listeria*, *Yersinia*, *Pasturella*, toxigenic *E. coli*. Obviously not all spread by eating the bird.

Ham: *Campylobacter*, *Listeria*, and pork (not ham) can have Toxoplasmosis and Hepatitis E.

Potatoes: botulism and salmonella.

Vegetables of all kinds: samonella.

Whipping Cream: *Bacillus*, salmonella, and toxigenic *E. coli.*

Canned goods: botulism.

Lettuce: salmonella, *Shigella*, *E. coli, C. difficile, and E. histolytica.*

Cranberries will not help the UTI you will get this weekend; the data is reasonably definitive that it has no utility in the treatment or prevention of cystitis.

The drunks who get aspiration pneumonia, and the carving trauma that leads to cellulitis.

There is the person making the meal. I had a patient on proton pump inhibitors, which increases the risk of *C. difficile* at least twofold, whose relative was recovering from *C. difficil*e while making the Thanksgiving meal. My patient developed moderately severe *C. difficile* diarrhea.

There is travel: influenza, tuberculosis, and other respiratory viruses acquired on airplanes and in airports.

Beware the Transportation Security Administration:

"it was noted that the agents wear the same gloves to pat down dozens, perhaps hundreds, of passengers. . . ."

Ewwwww. So make sure that you ask for new gloves when you get that cavity search, because you really do not know where those hands have been. Although, in fairness to the TSA, there have been no documented spread of infections.

Have a nice Thanksgiving, everyone. I know I will.

Rationalization

Grunkemeyer, V. L. Zoonoses, public health, and the backyard poultry flock. The veterinary clinics of North America. Exotic animal practice 14, 477-490, vi (2011).

Kenyon, T. A., Valway, S.E., Ihle, W.W., Onorato, I.M. & Castro, K.G. Transmission of multidrug-resistant Mycobacterium tuberculosis during a long airplane flight. The New England journal of medicine 334, 933-938 (1996).

Ooi, P. L. et al. Clinical and molecular evidence for transmission of novel influenza A(H1N1/2009) on a commercial airplane. Archives of internal medicine 170, 913-915 (2010).

Spreadin' the Glove: TSA Infecting U.S.? Latex coverings "have been in crotches, armpits, touching people who may be ill." Bob Unruh, WND, November 22, 2010.

Poll Results

At Thanksgiving I most worry about

- salmonella from the turkey. 6%
- influenza from the travel. 5%
- the drunken relative I suffer through once a year. 16%
- nothing. That is what wine and Valium are for. 59%
- meat is murder so I eat tofu (*Yersinia*). 11%
- Other Answers 3%

    -the sad fact my hosts put their own forks into every dish.

I Am Endeavoring, Ma'am, to Examine the Immune System Using Stone Knives and Bear Skins

I HAVE this standard line I use with persistent infections: unlike love, "X" is forever. "X" may be HSV or HIV or MRSA or the infection of the day. Many infections are not cured, only controlled, and in the case of herpes (HSV), poorly controlled.

About 20% of Americans are seropositive for HSV. Most have been asymptomatic and all have intermittent low-grade replication and shedding. Given the avidity with which humans use their lips and genitals, it is no surprise in the widespread prevalence of the disease. Knowledge of the historical epidemiology of STDs suggests that monogamy was as common in the past as today.

I may have told this story before, but years ago a colleague (really, not me) was at our national meeting in Las Vegas having a drink at the hotel bar, waiting for a ride, when he was propositioned for a "date." He declined and was asked, "What is this meeting? Business has not been this slow in years." I don't know if he explained that it was the Infectious Diseases Society of America annual conference, and that ID docs know the real motto is, "What happens in Vegas is transmitted to your significant other" and act accordingly.

Some unfortunate HSV patients have recurrent poorly healing, painful ulcers. These usually respond to suppressive acyclovir, but not in this patient. He has recurrent herpes that does not respond to suppressive antivirals, and so is sent to me.

His history maybe suggests recurrent upper respiratory infections and sinusitis, but it's not impressive, and there is no suggestion of classic opportunistic infections or patterns that suggest an immunodeficiency. His HIV test is negative, but I check his CD4 cell count anyway—normal. I also send in tests for HSV susceptibility (pending for weeks; I think it is sent by real, not metaphorical, snail mail) and quantitative immunoglobulins.

His IgG is half normal and he has barely measurable IgG2. Argh. Immunoglobulin levels correlate poorly with the disease, so are the low IgG2 levels the reason for the refractory HSV? Although there are a few case reports of odd HSV infections in people with CVID, given that clinical recurrence is often the hallmark of HSV, I can't tell that anyone has looked at HSV as a marker of CVID. Zoster is on the list, but not Simplex.

I found one article to suggest that no one with HSV makes IgG2, and another that IgG2 is made in those with relapsing diseases. Maybe I should just stop measuring antibody levels. I find a lot of low immunoglobulin levels, but remain uncertain if they are meaningful and if I should give intravenous immunoglobulin in these cases. It at least provides the patient with a possible partial explanation, so perhaps it is useful. My ability to examine the immune system is primitive at best; I might as well try to construct a mnemonic memory circuit.

I will try the immunoglobulin and see what happens.

Postscript

The patient, as is so often the case, changed insurance and was lost to follow-up.

Rationalization

Oksenhendler, E. et al. Infections in 252 patients with common variable immunodeficiency. Clinical infectious diseases : an official publication of the Infectious Diseases Society of America 46, 1547-1554 (2008).

Coleman, R. M. et al. IgG subclass antibodies to herpes simplex virus. The Journal of infectious diseases 151, 929-936 (1985).

Sundqvist, V.A., Linde, A. & Wahren, B. Virus-specific immunoglobulin G subclasses in herpes simplex and varicella-zoster virus infections. Journal of clinical microbiology 20, 94-98 (1984).

Bear Skins and Stone Knives: https://www.youtube.com/watch?v=F226oW-BHvvI

Poll Results

I prefer

- *Star Trek*, original series. 24%
- *Star Trek, Next Generation*. 41%
- *Star Trek, Deep Space* 9. 8%
- *Star Trek*: *Voyager*. 3%
- huh? What does this question have to do with the case? 21%
- Other Answers 3%

  -Star Wars! May the Force Be With You.

A Repeat after Twenty-nine Years

TWENTY-NINE years ago, as an afterthought, I did an ID rotation as a fourth-year medical student. It was, safe to say, a life-changing month, and set me on the road to what I am today. Fill in your own comment on just what it is that I am today.

I remember two cases as if they were yesterday—which is odd as I rarely remember yesterday. One sign of aging is that the past is easier to recall than the present. One was a mitral valve endocarditis that acutely ruptured and went to emergency valve surgery. The pathology was fascinating, a large mass of infection on the valve. It was cool. I have seen dozens of similar cases in the intervening years, and I still find endocarditis fascinating.

The other case was dreadful, an elderly male who developed a *Pseudomonas* urinary tract infection after a transurethral resection of the prostate. He developed bacteremia and meningitis. Ticarcillin and tobramycin were the only halfway active agents we had back then, and as best as I could tell, they had just enough activity against the organism to ensure that his death was slow. At autopsy, the brain was covered with a thick green slime of pus and *Pseudomonas*. All these years later and I have taken care of few cases that were as awful, yet in a horrible way, cool.

Until last month I had not seen another Pseudomonal meningitis, and I did not feel I was missing anything.

The only difference between the patient then and the one now are the improved antibiotics: we have agents that both cross the blood-brain barrier and can kill *Pseudomonas*. The infection was clinically cured, although whether he will wake up is uncertain. Even under the best of circumstances, bacterial meningitis is not associated with a return to normal cerebral function, and this is not the best of situations.

Predominantly a nosocomial infection, *Pseudomonas* meningitis has been reported to be community-acquired in a smattering of cases, as there is a smattering of everything reported on PubMed. Like most patients with gram-negative meningitis, the patient was elderly and had a source: in the first case, he had a UTI and bacteremia.

My patient this time had an intrathecal pain pump, five years old, that was causing him no other problems. Was this the source of central nervous system bacterial seeding? Hard to say, since the reported cases of meningitis occur around the time of pump placement. But one wonders.

And should we have worried about *Pseudomonas*? Probably not; he had no risk factors.

> *"P. aeruginosa caused 6.8% of 4114 unique patient episodes of GNR bacteremia upon hospital admission (incidence ratio, 5 cases per 10,000 hospital admissions). Independent predictors of P. aeruginosa bacteremia were severe immunodeficiency, age >90 years, receipt of antimicrobial therapy within past 30 days, and presence of a central venous catheter or a urinary device. Among 250 patients without severe immunodeficiency, if no predictor variables existed, the likelihood of having P. aeruginosa bacteremia was 1:42. If >or= 2 predictors existed, the risk increased to nearly 1:3."*

I hope to go at least another twenty-nine years before I see my next case—although at eighty-three I should be retired if not dead.

Postscript

The patient did not have a full central nervous system recovery and died in the nursing home.

Rationalization

Wunderlich, C. A. & Krach, L. E. Gram-negative meningitis and infections in individuals treated with intrathecal baclofen for spasticity: a retrospective study. Developmental medicine and child neurology 48, 450-455 (2006).

Schechner, V. et al. Gram-negative bacteremia upon hospital admission: when should Pseudomonas aeruginosa be suspected? Clinical infectious diseases : an official publication of the Infectious Diseases Society of America 48, 580-586 (2009).

Archenemy

THIS was a call weekend, and most of the consults were *S. aureus* infections in one form or another. If I were a superhero, then my archenemy would be *S. aureus.* In that case, it would probably be called Professor S. aureus or Doctor S. aureus or some other unearned title. Just for once I would like to see the PhD thesis of a supervillain.

One was an elderly male with three days of low-grade fever (100.2, not really a fever) followed by two days of temperature of 103 and severe back pain. The review of systems and the physical exam are negative; the initial workup shows some pyuria (pus in the urine). UTI is the leading diagnosis, and he is treated accordingly. However, urine and blood both grow MSSA. The MRI of his back is negative.

A couple of rules concerning MSSA in the blood.

Zero, low back pain without discitis or epidural abscess is a frequent manifestation of *S. aureus* bacteremia. I remember it happens 30% of the time, a reference-free "fact" I learned as a fellow.

First, if *S. aureus* is in the blood for no apparent reason, then the diagnosis is endocarditis. Period. He had a negative transthoracic echocardiogram, but it is no surprise. He is a large man, and it takes time to form a vegetation. I was told as a fellow late last century that it takes more than a week to grow a vegetation that can be seen on an ECHO. And, as I remember it, it was in a sheep model of endocarditis, although I have no reference to support this "fact" either. Still, that's my story and I am sticking to it.

Third, unlike other organisms, when *S. aureus* is in the blood and in the urine for no good reason (such as a Foley catheter), assume that the order is blood to urine, not urine to blood as is the case with virtually every other organism.

Although the clinical data, like all clinical data, has a variable association between *S. aureus* bacteremia and urinary tract seeding:

> *"In this retrospective cohort study, patients who had Staphylococcus aureus bacteremia but who lacked signs and symptoms of urinary tract infection due to S. aureus and who did not have an indwelling urinary catheter had a likelihood of S. aureus bacteriuria of 2.5% (2 of 79 patients). Therefore, we strongly question the theory that S. aureus bacteremia causes S. aureus bacteriuria."*

The real take-home with *S. aureus* in the blood is to ask, "Where did it come from and where did it go?" And above all, don't give a few days of IV antibiotics and then change to oral therapy. Down that path lies disaster. There are at least three studies that demonstrate improved outcomes for *S. aureus* bacteremia when an ID consult is obtained (he mentions self-servingly). Of course, those studies were all done by ID docs, but if you can't trust an ID doc, who can you trust?

Rationalization

Lee, B. K., Crossley, K. & Gerding, D. N. The association between Staphylococcus aureus bacteremia and bacteriuria. The American journal of medicine 65, 303-306 (1978).

Baraboutis, I.G. et al. Primary Staphylococcus aureus urinary tract infection: the role of undetected hematogenous seeding of the urinary tract. European journal of clinical microbiology & infectious diseases : official publication of the European Society of Clinical Microbiology 29, 1095-1101 (2010).

Ekkelenkamp, M.B., Verhoef, J. & Bonten, M.J. Quantifying the relationship between Staphylococcus aureus bacteremia and S. aureus bacteriuria: a retrospective analysis in a tertiary care hospital. Clinical infectious diseases : an official publication of the Infectious Diseases Society of America 44, 1457-1459 (2007).

Poll Results

I most trust

- clergy. 0%
- pharmacists. 13%
- nurses. 13%
- ID docs .26%
- no one. 38%
- Other Answers 9%
    - -myself.
    - -my wife.
    - -God.
    - -morticians.
    - -nitric oxide.
    - -Mom.

Starting a New Pentad

THERE are bugs that are only rarely seen in practice. Like the HACEK group. An acronym beloved by ID docs, it stands for *Haemophilus parainfluenzae, Haemophilus paraphrophilus, Aggregatibacter actinomycetemcomitans, Aggregatibacter aphrophilus, Cardiobacterium hominus, Eikenella corrodens, and Kingella kingii.*

They are gram-negative coccobacilli of the mouth, and usually are a cause of endocarditis, of which I have seen exactly one of each in twenty-five years. It took fifteen years to complete the pentad, and I have not seen one in the blood this century. Until, of course, this week.

The patient is a diabetic who comes in with a cough, mostly non-productive, but the CT scan showed multiple right -sided lung nodules. The nodules are both peripheral and central, and radiology is calling them more mass-like than consolidative, with some pneumonia as well.

The blood cultures and sputum are growing both MRSA and *Eikenella*. A review of systems reveals no reason for pneumonia, aspiration or otherwise, and the patient has a total of just four teeth, all in good condition. The echocardiogram is negative, so the bacteremia appears to be from the lung.

How to put it together? I bet the pulmonary masses were tumor, the patient being a former smoker, and that this was more of a post-obstructive pneumonia. The lack of teeth, risks, and the *Eikenella* made a primary pneumonia unlikely.

I was, as is often the case, wrong. The bronchoscopy showed no cancer and the repeat CT scan showed progressive multifocal, necrotic pneumonia.

Wrong again. One of my colleagues let me know that the pulmonologist particularly enjoyed my being wrong, since she never thought it was a malignancy. Sigh.

The PubMeds reveal a scant twenty-two cases of *Eikenella* pneumonia, including one infection in a Barbary ape that also had *S. aureus.* Most patients had aspiration as a reason, and this patient had no reason for the organism in his lower airway, as happens about 20% of the time. Necrosis is the rule with the organism, and he is slowly improving.

Part of the fun of writing these cases down is the odd places a PubMed search can take me. I wondered exactly where in the mouth *Eikenella* is found (subgingival, if you want to know), and wandered across a reference where the mouth flora was followed in patients on a Stone Age diet:

> *"Ten subjects living in an environment replicating the Stone Age for 4 weeks were enrolled in this study. Bleeding on probing (BOP), gingival and plaque indices, and probing depth (PD) were assessed at baseline and at 4 weeks. Microbiologic samples were collected at the mesio-buccal subgingival aspects of all teeth and from the dorsum of the tongue and were processed by checkerboard DNA-DNA hybridization methods.*
> *RESULTS: No subject had periodontitis. Mean BOP decreased from 34.8% to 12.6% (P <0.001). Mean gingival index scores changed from 0.38 to 0.43 (not statistically significant) and mean plaque scores increased from 0.68 to 1.47 (P <0.001). PD at sites of subgingival sampling decreased (mean difference: 0.2 mm; P <0.001). At week 4, the total bacterial count was higher*

> *(P <0.001) for 24 of 74 species, including Bacteroides ureolyticus, Eikenella corrodens, Lactobacillus acidophilus, Capnocytophaga ochracea, Escherichia coli, Fusobacterium nucleatum naviforme, Haemophilus influenzae, Helicobacter pylori, Porphyromonas endodontalis, Staphylococcus aureus (two strains), Streptococcus agalactiae, Streptococcus anginosis, and Streptococcus mitis"*

Since beer dates to the Neolithic, I suppose that drunken Stone Agers were at particular risk for aspiration pneumonia and that *Eikenella* was one of the pathogens. Although according to a report in *Discover*, getting drunk on ancient beer was no mean feat.

> *"It was unspeakable. To call it swill would be an insult to bad alcohol everywhere. Its angry, vinegary bouquet recalled descriptions of pruno, the prison hooch made of canned fruit cocktail, Wonder Bread, and ketchup."*

Although to be fair, the Neolithics had a bit more practice in brewing the beer that made the Stone Age famous. I'll stick with a West Coast IPA, thank you very much.

Rationalization

Brack, M. Eikenella corrodens-caused botryomycosis-type pneumonia in a barbary ape (Macaca sylvanus). APMIS : acta pathologica, microbiologica, et immunologica Scandinavica 105, 457-462 (1997).

Hoyler, S.L. & Antony, S. Eikenella corrodens: an unusual cause of severe parapneumonic infection and empyema in immunocompetent patients. Journal of the National Medical Association 93, 224-229 (2001).

Baumgartner, S. et al. The impact of the stone age diet on gingival conditions in the absence of oral hygiene. Journal of periodontology 80, 759-768 (2009).

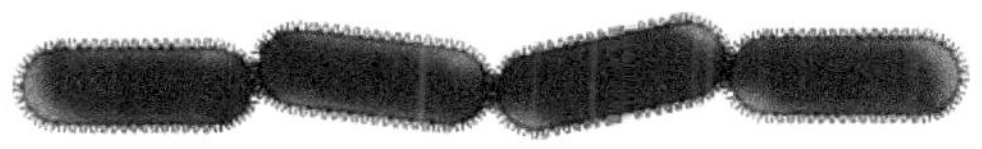

235 Days Later

No zombies. 235 days is what it took to drop about 45 pounds. Finally got rid of the steroid-induced middle-aged bloat. Losing weight is simple: fewer calories in than out and the weight comes off. Implementation is not so easy; it has been 235 days without a maple bar. Yeah, I know, First World problem. Boo hoo for me.

As I have mentioned many times, there are three reasons for all medical problems: genetics, wear and tear, and infectious, and only one is of any real interest. What do bacteria have to do with obesity? It turns out that there is an association between the makeup of the colon flora and weight gain.

There is a hodgepodge of studies that show a relationship between the bacterial constituents of the colon and weight gain. Some bacteria appear to be weight promoting:

> *"Evidence suggests that the metabolic activities of the gut microbiota facilitate the extraction of calories from ingested dietary substances and help to store these calories in host adipose tissue for later use. Furthermore, the gut bacterial flora of obese mice and humans include fewer Bacteroidetes and correspondingly more Firmicutes than that of their lean counterparts, suggesting that differences in caloric extraction of ingested food substances may be due to the composition of the gut microbiota. Bacterial lipopolysaccharide derived from the intestinal microbiota may act as a triggering factor linking inflammation to high-fat diet-induced metabolic syndrome."*

Antibiotics can make a difference:

> *"The BMI increased significantly and strongly in vancomycin-plus-gentamycin-treated patients (mean [+/-SE] kg/m(2),*

> *+2.3 [0.9], $p = 0.03$), but not in controls or in patients treated with other antibiotics." perhaps due to "the role of the gut colonization by Lactobacillus sp, a microorganism intrinsically resistant to vancomycin, used as a growth promoter in animals, and found at a high concentration in the feces of obese patients."*

and

> *"in some studies, the modifications of gut microbiota induced by antibiotics, prebiotics and probiotics led to improved inflammatory activity in parallel to amelioration of insulin sensitivity and decreased adiposity. However, these effects were mainly observed in animal models."*

It is an interesting topic, more a curiosity than of clinical applicability. Of course, association is not causation, and getting fat is multifactorial. If you have the right colonic flora, stuffing your face with three deluxe cheeseburgers at Five Guys—which I saw last week—isn't going to maintain your ideal body weight.

Given that there are 10 to 100 times more bacteria in and on us than there are cells of us, it is no surprise they control us. We are little more than a barely sentient bacterial transport and feeding mechanism. You think you are in control, but between the bacteria and the New World Order, you are but pawns in the universe to creatures below you in the evolutionary chain and above you in the economic chain. Sad thing is, there are people who will think this is not the joke it is meant to be.

Everyone wants a magic diet to help weight loss, and most are disappointed to learn it was eating fewer calories than I expended. Gut bacteria are not that solution to weight loss, although since I have not had a colon for fifteen years, I wonder what my flora may be now and its interaction with my weight and serum lipids.

Like much of medicine, interesting to speculate upon, but of no practical utility. Interestingly, no one is yet selling probiotics specifically to aid in weight loss. Perhaps they cannot advertise probiotics that way.

Rationalization

DiBaise, J. K. et al. Gut microbiota and its possible relationship with obesity. Mayo Clinic proceedings 83, 460-469 (2008).

Thuny, F. et al. Vancomycin treatment of infective endocarditis is linked with recently acquired obesity. PloS one 5, e9074 (2010).

Esteve, E., Ricart, W. & Fernandez-Real, J.M. Gut microbiota interactions with obesity, insulin resistance and type 2 diabetes: did gut microbiote co-evolve with insulin resistance? Current opinion in clinical nutrition and metabolic care 14, 483-490 (2011).

Poll Results

There

- is no free will. 9%
- is only free will on trivial issues: Coke or Pepsi. 4%
- is free will, but it matters not as control is elsewhere. 15%
- is the fact that I do what my spouse tells me. Or else. 20%
- is the truth that by choosing an answer I demonstrate free will. Or do I just think that? 43%
- Other Answers 9%

    -It's not free. Sometimes, however, it's cheap.

    -is the ability to choose.

My My. M.I.?

I HAVE said many times that most men would have been dead in their thirties if not for their significant others hauling their sorry carcasses into the emergency room. So often when I ask a patient why they came into the hospital, they give a surly, "My wife/girlfriend made me." And as a result we live rather than die.

The patient had severe rigors and fevers shortly after finishing dialysis and decided to tough it out for a day or three. If you are over fifty, with diabetes and with renal failure, it is best not to tough out a fever of 103 and rigors; odds are you have bacteremia, and bacteremia is bad.

What drove the patient into the ER, besides the spouse, was shortness of breath that reminded him of a prior myocardial infarction which had resulted in coronary artery stents. Evaluation in the ER was negative for an obvious infectious source, and after studies were drawn, antibiotics were given.

By the next day, the blood is growing a gram-negative rod, his creatine kinase-MB is 650 IU/L, and the troponin is 36 ng/mL. Creatine kinase-MB is a pretty specific test for heart muscle damage, and values should be less than 25 IU/L. The troponin test can indicate heart attack, but only in the presence of other confirming factors, although his level is markedly high.

The electrocardiogram is not impressive for ischemia or for myocardial infarction (MI). I am as far from a cardiologist as is medically possible. I am so old I used to rule in MIs with lactate dehydrogense isoenzyme tests. CK-MBs and troponins became de rigueur long after I concerned myself with issues of cardiac ischemia. But everyone assures me the troponins are higher than would be expected for the CK-MB. It is the sepsis:

> *"Cardiac troponins are elevated in 85% of patients with sepsis in the absence of acute coronary syndrome."*

and is associated with increased mortality, especially in dialysis patients.

> *"This study showed that elevated cTnI levels were significantly associated with short- and long-term mortality in ESRD patients with sepsis. "*

While one of the pathophysiological bees in my bonnet is the idea that infection causes inflammation which is prothrombotic and is associated with both short- and long-term vascular events, the troponin bump in sepsis may not be due to a vascular thrombotic event.

> *"We found no differences in coagulation parameters analyzed with rotational thrombelastometry between cTnI-positive and -negative patients with SIRS, severe sepsis, and septic shock. These findings suggest that pathophysiological mechanisms other than thrombus-associated myocardial damage might play a major role, including reversible myocardial membrane leakage and/or cytokine mediated apoptosis in these patients."*

I never argue with the results of rotational thrombelastometry, or with any other modality I cannot spell, much less understand. So maybe the bump in troponins is a direct cytokine effect or endotoxin effect; both have been suggested. What to do about it is another question, since if the effect is not due to thrombosis, at least acutely, there would not appear to be any intervention to offer.

Rationalization

Williams, S. G. et al. Troponin T: how high is high? Relationship and differences between serum cardiac markers according to level of creatine kinase and type of myocardial infarction. Postgraduate medical journal 80, 613-614 (2004).

Smith, A., John, M., Trout, R., Davis, E. & Moningi, S. Elevated cardiac troponins in sepsis: what do they signify? The West Virginia medical journal 105, 29-32 (2009).

Altmann, D. R. et al. Elevated cardiac troponin I in sepsis and septic shock: no evidence for thrombus associated myocardial necrosis. PloS one 5, e9017 (2010).

Kang, E. W. et al. Prognostic value of elevated cardiac troponin I in ESRD patients with sepsis. Nephrology, dialysis, transplantation : official publication of the European Dialysis and Transplant Association - European Renal Association 24, 1568-1573 (2009).

Poll Results

I never argue with

- rotational thrombelastometry. 17%
- my wife when she says go to the ER. 10%
- my husband when he says go to the ER. 5%
- I argue constantly with everyone.
- I am fifteen years old. 29%
- success. 38%
- Other Answers 2%

    -nitric oxide.

Blood Cultures

BLOOD culture results come in four flavors. Figuratively speaking, unless they are a vampire's amuse-bouche.

First is real and important. *Candida, E. coli, S.aureus.* Try as you might want to, you can never, or almost never, declare some organisms a contaminant. Only once have I declared *S. aureus* a contaminant; it was in a blood culture that was drawn through a psoriatic plaque. Given that most, if not all, chronic skin conditions are awash in *S. aureus,* and otherwise there was no suggestion of infection (the blood cultures being done, evidently, just because) I said—with no little trepidation—"Ignore it." I was right, but have not pushed my luck, or expertise, since.

Second is real but unimportant. Given the ubiquity of bacteremia from the gastrointestincal tract, occasionally we pick up a 1 out of 4 *S. viridans.* The bacteremia is real, but not causing disease. You need to evaluate the patient carefully, but often it is nothing. It is my impression, neither confirmed nor denied by the literature (remember the three most dangerous words in medicine: in my experience), that patients who are found down for prolonged periods of time have real but unimportant positive blood cultures with coagulase-negative *Staphylococcus.* I figure prolonged pressure on the skin gives a chance for the bacteria to invade the bloodstream, or perhaps it is simply raging confirmation bias.

Third is contaminant: the 1 out of 4 coagulase-negative *Staphylococcus* or the occasional Diphtheroid, the skin bugs that only rarely get in the blood culture for real; most of the time they can be safely ignored. Most of the time. Cue the music: Jason Voorhees from *Friday the 13 th* is coming.

And then there are the, "Huh, what do I do now?" cultures. The patient is admitted with fevers and with an altered mental status while on dialysis. Not uncommon. After cultures, she is given a slug of vancomycin and pipercillin/tazobsactam, the polyantibiotics of choice for everyone but me. I have never gotten into the habit of the beta-lactamase inhibitor combinations, perhaps a story for another time.

The patient perks right up and after 3.5 days both sets of blood cultures have gram-positive rods that after 5 days are called *Corynebacterium striatum*. To make matters more complicated, she not only has a fistula for dialysis, but a five-year-old pacemaker and a painful shoulder that appears, upon further workup, to be chronic osteomyelitis of a former fracture site that was plated. The echocardiogram, you ask? Pristine.

So huh, what do I do now? No way is the *Corynebacterium* from the shoulder. You almost never get positive blood cultures from osteomyelitis. The ECHO result makes a pacer infection less likely and cardiology will (appropriately) not pull the pacer without more compelling data suggesting that it is the source of the bacteremia. There is no fluid around the fistula. The patient was ill on admit, suggesting the cultures were the cause, but perhaps it was a transient large bolus of bacteria. My colleague, new to the practice, was mentioning that he often gets called to answer questions that have no answers. I told him that being a consultant is to be ignorant with style and to make up answers with confidence. Welcome to ID.

C. striatum is a pathogen, infecting lungs, joints, lines, and pacers. Part of the issue is that the labs often do not speciate *Corynebacteria*, so the organism may be passed off as nothing or as normal flora. My plan? I pulled it out of . . . thin air. Let's treat with six weeks of Vancomycin, get the shoulder debrided (different doctor, different hospital), and if the bacteremia recurs, get the pacer pulled and the fistula reevaluated.

Postscript

None. Lost to follow-up due to insurance reasons and transfer of care.

Rationalization

Oliva, A. et al. Pacemaker lead endocarditis due to multidrug-resistant Corynebacterium striatum detected with sonication of the device. Journal of clinical microbiology 48, 4669-4671 (2010).

von Graevenitz, A., Frommelt, L., Punter-Streit, V. & Funke, G. Diversity of coryneforms found in infections following prosthetic joint insertion and open fractures. Infection 26, 36-38 (1998).

Fernandez-Roblas, R. et al. In vitro activity of tigecycline and 10 other antimicrobials against clinical isolates of the genus Corynebacterium. International journal of antimicrobial agents 33, 453-455 (2009).

You Heard It Here First

Some antibiotics are well-known to affect hearing: aminoglycosides, vancomycin, erythromycin, and quinine are the common ones. But cephalosporins?

The patient is an elderly male with his second liver abscess in ten years. I don't have a good reason for the liver abscesses. Both times it was a *S. anginosus*, but years apart. But he has been getting better on intravenous antibiotics.

He notes new, at least exacerbated, hearing problems. He says he gets roaring in his ears after the infusions, and when the roaring occurs, his hearing decreases. Could it be the antibiotics?

Initially I say no, but as I talk with him it turns out this has been a problem for years and was better since he was put on a low salt diet. "Do you have Ménière's disease?" I ask.

He is not certain. A few weeks ago I had another patient who had similar symptoms on the last week of therapy with ceftriaxone, and I told the patient it was probably not due to the ceftriaxone. But now I wonder.

Two grams of ceftriaxone has 7.2 milliequivalents of sodium, about 160 milligrams—or equal to a small bag of potato chips.

I wonder if the sudden salt load is making the Ménière's worse. As best I can tell, salt restriction improves the symptoms of Ménière's and salt loading increases the inner ear fluid production. The closest I can find is worsening of Ménière's with salt loading and with PMS. I can't find reports of worsening of Ménière's symptoms with salt loading from treating patients, say with intravenous fluids, or other reports with IV antibiotics. It seems plausible.

There is a study in there for someone, and when you do it, give me the credit. As I have said before, my main professional goal is to have something in medicine named after me, so we will henceforth call this the Crislip Effect: exacerbation of Ménière's due to antibiotic-induced salt load. Or someone with better Google-Fu will find the prior report that I can't.

Rationalization

Ménière's Disease: http://emedicine.medscape.com/article/1159069-overview

ten Cate, W J., Curtis, L. M. & Rarey, K. E. Effects of low-sodium, high-potassium dietary intake on cochlear lateral wall Na+,K(+)-ATPase. European archives of oto-rhino-laryngology : official journal of the European Federation of Oto-Rhino-Laryngological Societies 251, 6-11 (1994).

Silverstein, H. & Takeda, T. Sodium loading of inner ear fluids. The Annals of otology, rhinology, and laryngology 85, 769-775 (1976).

Andrews, J.C., Ator, G.A. & Honrubia, V. The exacerbation of symptoms in Meniere's disease during the premenstrual period. Archives of otolaryngology--head & neck surgery 118, 74-78 (1992).

Poll Results

My goal in medicine is to

- have something named after me. 11%
- have a Medicare free practice. 8%
- avoid dying of a disease in my specialty. 31%
- retire without divorce or substance abuse. 39%
- write a Medscape blog. 10%
- Other Answers 2%

Tis the Season for ID

THE week leading up to Christmas is always a slow one. I am not entirely certain why, since the bugs don't know it is a holiday. But every year my practice gets quiet until after the New Year, a phenomenon not isolated to infectious diseases.

I previously ran through all of the infections associated with the Thanksgiving calorie fest, but I could not see where Christmas has any risks or associations with infectious diseases. So, for hoots and giggles, I entered,"Christmas" and "Infection" into the PubMeds. Ninety-for hits! Oh, wait. Christmas is the name of several authors. But there are some Christmas-associated infections.

Metarhizium and *Beauveria* spp. can be isolated from Christmas trees in my home of the Willamette Valley in Oregon, but are not human pathogens. *Cryptococcus gattii* is found in high numbers in Pacific Northwest firs, so I wonder if there will be a slight uptick in *Cryptococcus* cases following the indoor tree season. You heard it here first. Doesn't seem worth getting a fake tree, however.

The rash of secondary syphilis can have a Christmas tree pattern, so be careful who you stand under the mistletoe with, especially since

> *The trends point consistently to an increase in sexual activity and unsafe sex occurring at or around the Christmas period.*

Time for Christmas parties, so crowding and sharing of food could have the potential to amplify infectious diseases, like influenza and norovirus:

Thirty-five of 61 individuals interviewed (attack rate, 57.4%) fell ill. In 94.3% of cases, the onset of illness was within 48 h of a Christmas party at the facility.

and typhoid fever, which according to an abstract, appears to have been an early form of biologic warfare.

In 1941, at Christmas time more than 600 German soldiers visited Paris and were infected with the bacillus of Eberth, the agent of typhoid fever. All of the infections occurred at the Brasserie La Brune situated in the center of Paris. It was, in fact, an act of Resistance.

There have been *Trichinella spiralis* and salmonella outbreaks, so be careful what you eat, especially if you are part of an occupying army. My theory, not yet proven, is if you get the blood alcohol levels high enough it will kill any infection. And perhaps you as well.

Drugs can get into the Christmas spirit:

The patient was a 78-year-old man admitted for induction of chemotherapy for acute myelogenous leukemia (AML) who began to have auditory hallucinations, specifically of Christmas music, the 2nd day of voriconazole therapy. His psychiatric evaluation was otherwise unremarkable. After discontinuing voriconazole the hallucinations decreased in intensity by the 2nd day and ceased altogether by the 3rd day.

That could drive me to suicide. Rup a pum pum induced death, although my least favorite Christmas song is easily *Feliz Navidad*. They know that in the clinic and make a point of playing it for me.

Even fictional characters do not get spared:

Every Christmas we sing about Rudolph the red-nosed Reindeer, but do we give much thought to why his nose is red? The general consensus is that Rudolf has caught a cold, but as far as I know no proper diagnosis has been made of his abnormal condition. I think that, rather than having a cold, Rudolf is suffering from a parasitic infection of his respiratory system. To some this may seem a bit far-fetched as one would not expect an animal living with Santa Claus at the North Pole to be plagued by parasites, but I shall show otherwise.

Reindeer is our traditional Christmas dinner, and rabbit is what we eat for Easter. It helped the kids understand the lack of presents and chocolate eggs.

Hanukkah is infection free, as least as a PubMed search term. Whether a cultural reporting bias or intrinsically a safer celebration, I cannot say.

Thanks to all who read this blog, I wish you all a generic non-denominational seasonal greeting. May your holidays be filled with germs.

Rationalization

Keatinge, W.R. & Donaldson, G.C. Changes in mortalities and hospital admissions associated with holidays and respiratory illness: implications for medical services. Journal of evaluation in clinical practice 11, 275-281 (2005).

Wellings, K., Macdowall, W., Catchpole, M. & Goodrich, J. Seasonal variations in sexual activity and their implications for sexual health promotion. Journal of the Royal Society of Medicine 92, 60-64 (1999).

Wollenberg, A. & Eames, T. Skin diseases following a Christmas tree pattern. Clinics in dermatology 29, 189-194 (2011).

Medici, M.C. et al. An outbreak of norovirus infection in an Italian residential-care facility for the elderly. Clinical microbiology and infection : the official publication of the European Society of Clinical Microbiology and Infectious Diseases 15, 97-100 (2009).

Brumpt, L., Petithory, J.C. & Ardoin, F. [An epidemic of typhoid fever among the German troops in Paris, Christmas 1941]. Histoire des sciences medicales 42, 17-20 (2008).

Agrawal, A.K. & Sherman, L.K. Voriconazole-induced musical hallucinations. Infection 32, 293-295 (2004).

Haim, M. et al. An outbreak of Trichinella spiralis infection in southern Lebanon. Epidemiology and infection 119, 357-362 (1997).

Roels, T.H. et al. Incomplete sanitation of a meat grinder and ingestion of raw ground beef: contributing factors to a large outbreak of Salmonella typhimurium infection. Epidemiology and infection 119, 127-134 (1997).

Halvorsen, O. Epidemiology of reindeer parasites. Parasitology today 2, 334-339 (1986).

Poll Results

- I would like to wish one and all
- Merry Christmas. 50%
- Happy Hanukkah. 4%
- Joyful Kwanza. 0%
- Happy Festivis. 12%
- Drunken Newtonmas. 19%
- Other Answers 15%
  - -HAPPY HOLIDAYS
  - -a pleasant solstice.

Jawing Away

NEXT weekend is the New Year's weekend, but I am on call, so it is not going to be a holiday for me. Might even miss the Ducks' triumphant Rose Bowl victory—such is the price of call.

Work is picking up as well. It is always slow the week before Christmas, as though the bugs know it is the holidays. It has been a good thing it has been slow, as I have been ill and the interferon-induced mental haze is conducive to neither doctoring nor writing. This afternoon is the first time in a week I have felt halfway normal, or at least what passes for normal in the health care environment.

I saw a patient with a history of a squamous cell carcinoma of the cheek. Despite resection and no bone involvement, she had increasing pain in the jaw. She had debridement and a few teeth pulled that were not doing well. Prior to this extraction, besides advanced gingivitis, she had no trauma to the jaw, radiation, chemotherapy, or other extractions.

After the extractions, the pain in the jaw continued and she was sent to me when the pathology showed inflammation, osteomyelitis, and organisms that looked like actinomycosis. Unfortunately, there were no cultures sent, but *Actinomyces* look reasonably characteristic, branching gram-postive rods.

This was not lumpy jaw, a classic manifestation of *Actinomyces*, but what appeared to be a primary odontogenic jaw osteomyelitis due to actinomycosis. So why was that?

Well, she had been on bisphosphonates for her osteoporosis. There is an association (which does not equal causation) between osteonecrosis of the jaw from bisphosphonates and actinomycosis:

> *"The etiology of bisphosphonate-related osteonecrosis of the jaw is unknown but was initially postulated to be mediated by bisphosphonate accumulation within the jaws, resulting in avascular necrosis. Bisphosphonates may not be the primary cause. Actinomyces are an under-recognized agent in pathogenesis, and timely actinomycosis-specific treatment may improve outcome."*

and

> *"Actinomycosis of the jaws is a rare disease, which has been recently described in patients with infected osteoradionecrosis (IORN) and bisphosphonate-associated osteonecrosis (BON). We investigated our archive material for Actinomycosis of the jaws with special regard to underlying disease. Out of a total number of 45 patients with Actinomycosis, 43 (93.5%) suffered from BON (58.7%) or IORN (35.6%)."*

as examples.

You do not have to have the more classic osteonecrosis to get the bisphosphonate-jaw associated osteomyelitis from actinomycosis, although it is a much rarer manifestation. Why and if actinomycosis causes, or is a complication of, biphsophonates is not known.

It will be interesting to see how her jaw does on treatment. Every disease has at its heart an infectious etiology. Or at least the interesting ones do.

Postscript

The infection responded to a long course of penicillin.

Rationalization

Naik, N.H. & Russo, T.A. Bisphosphonate-related osteonecrosis of the jaw: the role of actinomyces. Clinical infectious diseases : an official publication of the Infectious Diseases Society of America 49, 1729-1732 (2009).

Hansen, T. et al. Actinomycosis of the jaws--histopathological study of 45 patients shows significant involvement in bisphosphonate-associated osteonecrosis and infected osteoradionecrosis. Virchows Archiv : an international journal of pathology 451, 1009-1017 (2007).

Premature Closure

It is something we all suffer from, and I find it more of a problem as I get older. In my own case, I tend not to think beyond the infectious disease differential diagnosis. Not everything, unfortunately, is due to an infection, and I occasionally have to force myself to consider diseases outside of ID.

There are other forms of premature closure, or at least of not considering diseases outside a narrow differential. It happens all the time when patients return from overseas travel. We think that the infection must be due to a disease found in the country that the patient visited. The physician is not considering that the presentation may be due to a more mundane infection and acquired closer to home.

Today's patient became ill with fevers to 104 eight days after returning from a trip to Africa. He had fever, a mild headache, diarrhea, and then a trunk and arm macular-papular rash after being admitted to the hospital.

Nothing on exam except the rash. The labs were significant for a white cell count of 2,000 per microliter, or about half the lower limit of normal, with a lymphopenia (particularly low lymphocyte count). His platelets were also low: 95,000 cells per microliter, when 150,000 is the lower limit of normal. There was also a mild anemia. Urinalysis, chest x-ray, and a chemistry panel were normal. The malaria smear, read by the night shift, was read as negative.

The intern calls me that night asking me what it could be. She reports that the patient was careful on the trip and only had no dietary indiscretions or exposures. The big four, I say, are yellow fever, dengue, malaria, and HIV, and with those labs it is probably HIV. Dengue and yellow fever usually give a transaminitis, as do the *Rickettsia* and many of the viral infections that drop the platelets. I'll see him in the morning.

The next morning the patient did not look all that sick, and exposure history was negative except for safe sex with men, both here and on his trip.

He also had Faget's sign—pulse-temperature disassociation. For every degree of temperature elevation, the pulse should go up about 10 beats per minute; but for a temp of 104, his pulse was only 69. Classically, Faget's is due to *Brucella*, typhoid fever, tularemia, and yellow fever; but all four should have liver involvement, and the history was not impressive for a good exposure. The Faget's threw me, and the lack of adenopathy and sore throat went against HIV.

The diarrhea I ignored once the preliminary stool study was negative, as I tend to do. Fevers and diarrhea go together like cookies and milk. The same cytokines that cause fever lead to loose stools. My just-so story is that in the old days, many infections were acquired from contaminated food and water so rapid elimination by vomiting and diarrhea is an important evolved response. That's my just-so story anyway.

An HIV antibody test was negative, so I figured let's check an HIV viral load, since if this is HIV it is acute disease. In acute disease the antibodies haven't had a chance to develop and so the test would be negative. I also did a VDRL (venereal disease research laboratory test for syphilis, although the rash spared the hands and feet so it wasn't characteristic of that. Neither HIV nor syphilis has been associated with Faget's, so I was less certain that it was either disease. It is too early to get serology for the odd infections; we can check for *Brucella* and other diseases in a couple of weeks.

And it was acute HIV. The viral load was positive. It could have been acquired overseas or at home; there is no way to know for certain. Common things are common, hence the tautology.

Rationalization

Souza, L.J. et al. Aminotransferase changes and acute hepatitis in patients with dengue fever: analysis of 1,585 cases. The Brazilian journal of infectious diseases : an official publication of the Brazilian Society of Infectious Diseases 8, 156-163 (2004).

The Girl with Faget's Sign. Mark Crislip, MD http://www.medscape.com/viewarticle/729248

Poll Results

2011 is over. Happy New Year.

- I am hoping for lots of odd infections. 17%
- I dread 2012. The world will end. So the Mayans predict. 5%
- I embrace 2012. The world will end. So the Democrats'/Republicans'predict based on the election results. 14%
- I can't believe we are ten years into the 2000s. Where did the last century go? I still write 1911 on my checks. 24%
- In a mere ten years we will see the first med students born this century. How did I get so old? 37%
- Other Answers 3%

  -optimistic.

  -I am an optimist. All will be well.

It Bugs Me Not to Have an Answer

I LIKE to have answers. More often than not, patients get infections for no good reason. Random bad luck, if luck exists, and the next thing you know, you are septic with meningococcus. I like reasons for infections, and while I cannot always come up with an explanation, it doesn't stop me from trying.

The patient is young and healthy. No risks for any infection. He helped a friend move a couch (purchased on Craigslist) and he spent the night on said couch.

He woke up with a half a dozen bites that he thought nothing of at the time, but over the next several days one bite, if bite it was, became increasingly red and tender. He treated himself with ibuprofen, but he developed fevers, rigors, and vomiting and thence to the ER.

He had an early necrotizing fasciitis treated with debridement. Cultures grew MRSA. During the time he received a call from a friend to let him know that he had an exterminator over and that the couch had bedbugs, or perhaps couchbugs. When I saw him, he had what could have been healing bed bug bites. Or just MRSA folliculitis.

So why the infection?

Well, MRSA is perhaps the most common cause of necrotizing fasciitis these days:

> *"In 247 cases, 42 microbial species were identified. S. aureus was the major prevalent pathogen and MRSA accounted for 19.8% of NF cases. Most patients had many coexisting medical conditions, including diabetes mellitus, followed by hypertension, chronic azotemia and chronic hepatic disease in order of decreasing prevalence."*

Although the Taiwan strain of MRSA is different from the U.S.A. strain.

Usually the association is between Group A streptococcus, NSAIDS like ibuprofen), and kids, but I am increasingly suspicious that inflammation is good, and you anti it at your risk.

Lastly, I blame the bedbugs:

> *The phenotype of the MRSA recovered from the bedbugs was consistent with community-associated MRSA and identical to that found on antibiograms from patients with MRSA infection who reside in this community. Given the high prevalence of MRSA (particularly USA300) in hotels and rooming houses in Vancouver's Downtown Eastside, bedbugs may become colonized with community-associated MRSA. Consequently, these insects may act as a hidden environmental reservoir for MRSA and may promote the spread of MRSA in impoverished and overcrowded communities.*

The infection was temporally related to the bites and he had none of the usual risks for MRSA.

And if you see free furniture on the roadside, remember there may be a reason besides aesthetics why it is being discarded.

Rationalization

Souyri, C., Olivier. P., Grolleau, S., Lapeyre-Mestre, M. & French Network of Pharmacovigilance, C. Severe necrotizing soft-tissue infections and non-steroidal anti-inflammatory drugs. Clinical and experimental dermatology 33, 249-255 (2008).

Lowe, C. F. & Romney, M.G. Bedbugs as Vectors for Drug-Resistant Bacteria. Emerging infectious diseases 17, 1132-1134 (2011).

Changchien, C.H. et al. Retrospective study of necrotizing fasciitis and characterization of its associated methicillin-resistant Staphylococcus aureus in Taiwan. BMC infectious diseases 11, 297 (2011).

Duration

How long? Not uncommon that the reason for consultation is the duration of therapy. Sometimes the answer is well established; sometimes it isn't. For some infections, like endocarditis, the answer is, "the longest course that will work without needing valve replacement." If 6 weeks of IV antibiotics doesn't cure your mitral valve endocarditis, then 8, 10, or 12 will not either.

For other infections, the end point is cure. I tend to treat a liver abscess until it is all gone.

Some infections have had a slowly decreasing number of days for the duration of therapy. In my career, the duration of antibiotics for acute sinusitis has gone from 14 to 10 to 7 to 3 to 0, all courses with the same outcome.

Most of the recommendations for duration of therapy are based on a combination of tradition and either the number of days in the week or the number of fingers on both hands. If I were to lose a hand, perhaps antibiotic use would decline by half. The Babylonian system of mathematics was sexagesimal (not as fun as it sounds, it is means based on 60) and probably resulted in excessive antibiotic duration, although they did originate the 7-day week.

Trust the French to use a different base for counting. I am a Francophile, but the French do love to be different. They compared 8 versus 15 days of antibiotics for ventilator-associated pneumonia and found no difference in outcomes. Why 8 days? And why not 16, double 8? Perhaps it is left over from the Roman Empire, which had an 8-day week for a time. Maybe it is left over from the Revolution, where they did all sorts of weird renumbering strategies, including de-incrementing by a head.

I was asked to see a very obese patient who had severe cellulitis with bacteremia from Group A *Streptococcus.* How long to treat? Reviews find no significant difference between different antibiotics in clinical cure after 30 days. The state of the art, for a disease we have been treating for 50 years with antibiotics, suggests,

> *We don't know whether antibiotics are as effective when given orally as when given intravenously, or whether intramuscular administration is more effective than intravenous. A 5-day course of antibiotics may be as effective as a 10-day course at curing the infection and preventing early recurrence.*

It amazes me that there have been no definitive cellulitis studies in all this time. Obese people, for whatever reason, do not respond as quickly to antibiotics, and my rule of thumb is to add a day to respond for every fifty extra pounds. I pulled that out of thin air, but it seems to work—he says with no confirmation bias, nope, not me.

But the bacteremia means we need to treat longer, right? Right? There are diseases where it is well defined that bacteremia calls for an increase in the duration of IV therapy, like *S. aureus*, *S. aureus* and, well, *S. aureus*. Most other diseases? Not so much. For many illnesses, just because they are bacteremic, it doesn't mean a longer course of therapy:

> *"Twenty-four eligible trials were identified, including one trial focusing exclusively on bacteremia, zero in catheter related bloodstream infection, three in intra-abdominal infection, six in pyelonephritis, 13 in pneumonia and one in skin and soft tissue infection. Thirteen studies reported on 227 patients with bacteremia allocated to shorter or longer durations of treatment.*

Outcome data were available for 155 bacteremic patients: neonatal bacteremia (n=66); intra- abdominal infection (40); pyelonephritis (9); and pneumonia (40). Among bacteremic patients receiving shorter (5-7 days) versus longer (7-21 days) antibiotic therapy, no significant difference was detected with respect to rates of clinical cure (45/52 versus 47/49, risk ratio 0.88, 95% confidence interval [CI] 0.77-1.01), microbiologic cure (28/28 versus 30/32, risk ratio 1.05, 95% CI 0.91-1.21), and survival (15/17 versus 26/29, risk ratio 0.97, 95% CI 0.76-1.23) ...No significant differences in clinical cure, microbiologic cure and survival were detected among bacteremic patients receiving shorter versus longer duration antibiotic therapy. An adequately powered randomized trial of bacteremic patients is needed to confirm these findings.

So I said the usual: give IV until they are better (afebrile, normal white cell count) and the leg looks good enough, then change to oral, for a total duration of how many fingers do you have. I hope I never give those suggestions to a one-armed doctor. I will be SO embarrassed.

Rationalization

Chastre, J. et al. Comparison of 8 vs 15 days of antibiotic therapy for ventilator-associated pneumonia in adults: a randomized trial. Jama 290, 2588-2598 (2003).

Morris, A.D. Cellulitis and erysipelas. BMJ clinical evidence 2008 (2008).

Havey, T.C., Fowler, R.A. & Daneman, N. Duration of antibiotic therapy for bacteremia: a systematic review and meta-analysis. Critical care 15, R267 (2011).

Poll Results

I base my antibiotic durations on

- the literature, mostly nineteenth- century poets. 31%
- ID recommendations. 21%
- a random number generator. 17%
- there is no rhyme or reason; I don't even read poets. 10%
- I ask Dr. Science! He Knows More than You Do! 17%

Rules

THERE are many rules of thumb in medicine. All rules, of course, have exceptions. Except for this rule. Captain Kirk could have used that to burn out the android Norman in "I, Mudd" instead of the "I am lying" paradox.

One of my rules is that the labs are used to support the clinical diagnosis, and if the labs do not, it is the labs, not the diagnosis, that are wrong. This rule is, as you can guess, not that reliable, and occasionally the labs will give the hint of the (possible) diagnosis.

The patient is on his third cycle of RVD for myeloma. RVD? Stands for lenalidomide, bortezomib, and dexamethasone. Why not LBD? Typical of oncology, where the initials have nothing to do with the generic drug, but the brand name, and not even consistently.

Within hours after the chemo, the patient has severe shortness of breath and a fever, comes to the ER, and is admitted quite hypoxic. No risk factors for unusual infections outside the immunosuppression and some bird exposure. All of the specimens fail to reveal a diagnosis, or at least an infectious diagnosis.

I order one lab that is abnormal: the lactate dehydrogenase is over 600 units per liter, when normal is 280 or below. The rest of the transaminases are normal. I miss the LDH as part of the normal comp. I don't miss the uric acid. Back in the day, everyone came in on allopurinol due to an elevated routine uric acid. "Don't ask, don't treat" is useful with uric acid. But I order a fair number of LDHs on odd pulmonary infiltrates.

Elevated LDH just indicates tissue damage in general, but because of his symptoms, I start thinking lung. Another rule of thumb is that there are two pulmonary processes that will give a markedly elevated LDH: PJP and drug lung, especially of the

BOOP (bronchiolitis obliterans with organizing pneumonia) variety. The evaluation for PJP is negative, so that leads to the question of drug lung, and then to the Googles. There is no doubt that PubMed and Google always help me look like I know more than I really do, but so often I seem to be the only one who bothers to search the interwebs.

Both lenalidomide and bortezomide cause lung disease: hypersensitivity pneumonitis and interstitial lung diseases. So this all may be drug-induced lung disease, not uncommon with cancer chemo agents.

An elevated LDH is nowhere near 100% in drug lung, more like 50%. Like many tests, given the operational parameters, it is more valuable when abnormal than when normal. But LDH may give a hint that, combined with a negative bronchoalveolar lavage, is suggestive of the diagnosis.

Short of a lung biopsy, there is no way to prove the diagnosis of drug lung, so I felt the need to continue some of the antimicrobials, especially for treatable atypical pneumonia.

But it was drug lung.

Rationalization

Yamamoto, M., Ina, Y., Kitaichi, M., Harasawa, M. & Tamura, M. Clinical features of BOOP in Japan. Chest 102, 21S-25S (1992).

Chen, Y., Kiatsimkul, P., Nugent, K. & Raj, R. Lenalidomide-induced interstitial lung disease. Pharmacotherapy 30, 325 (2010).

Lerch, E., Gyorik, S., Feilchenfeldt, J., Mazzucchelli, L. & Quadri, F. A case of lenalidomide-induced hypersensitivity pneumonitis. Onkologie 33, 249-252 (2010).

Thornburg, A., Abonour, R., Smith, P., Knox, K. & Twigg, H.L., 3rd Hypersensitivity pneumonitis-like syndrome associated with the use of lenalidomide. Chest 131, 1572-1574 (2007).

Gotoh, A. et al. [Lung injury associated with bortezomib therapy in relapsed/refractory multiple myeloma in Japan: a questionnaire-based report from the "Lung Injury by Bortezomib" Joint Committee of the Japanese Society of Hematology and the Japanese Society of Clinical Hematology]. [Rinsho ketsueki] The Japanese journal of clinical hematology 47, 1521-1527 (2006).

Poll Results

Rules to Guide Me Are

- The labs should support the history and physical, not the other way around. 24%
- The patient will be most talkative at five o'clock on Friday. 24%
- The less critical the phone call, the more likely it will be made at 2:00 a.m. 34%
- ID docs, above all others, like to hear the sound of their own voices because they are most often correct. 16%
- Always treat an abnormal lab. 0%
- Other Answers 3%

    -After a careful history and physical exam when you still have no diagnosis, ask the patient what their opinion is.

Avoiding Michael Faraday

The patient is an elderly female, mostly in good health, who has a headache, confusion, and dysarthria. She comes into the ER and has a negative CT scan, is afebrile, and is admitted to the stroke service. Shortly after admit, she has a fever and her mental status worsens.

The spinal tap shows elevated protein, 88 mg/dL when normal is 15-45. The cerebrospinal fluid (CSF) also contains 33,000 white blood cells per microliter, mostly lymphocytes, when the fluid should be crystal clear with no cells at all. She has some sort of meningitis—but whether it's viral or bacterial we can't tell yet. Another CT scan is negative, and pending lab studies acyclovir is started due to the worries of herpes simplex virus encephalitis.

Five days later, the third CT scan shows a large temporal-parietal defect in the brain; a polymerase chain reaction test for HSV in the spinal fluid is positive. Evidently, she had a distant history of cold sores, but nothing for years. HSV encephalitis is not a good disease.

CT scans are notoriously bad for finding early changes in HSV encephalitis. Unfortunately, a changing magnetic field leads to a current in a wire—electromagnetism 101 (and the basis of most of our electricity, at least here in the land of hydroelectric power). Why is that an important piece of electrical information? She has a defibrillator in, so the MRI—the diagnostic test of choice—is contraindicated. That is one wire you do not want to induce a current in. The defibrillator is there to prevent, not cause, arrhythmias. And many electronic devices do not do well when subjected to a burst of magnetism. Just ask Magneto, the mutant from *X-Men*.

No matter the alleged contraindication, there is always some poor patient who gets a study that in theory, they should not. Like lumbar punctures with mass-occupying lesions.

Do enough contraindicated studies and you can publish a series, as Nazarian et al. did at Johns Hopkins. A group there published a study of 555 MRIs on 438 patients (bet you can't have just one) with implanted cardiac devices, and what do you know? Nobody died and nobody had their device explode. Not bad.

They gave the caveat that not all devices have been "tested." I like that. Tested. In other words, one device may yet blow up or shock the unsuspecting patient. That is so nice. I am old enough to remember not only a time before MRIs but also CTs. The first time I saw an MRI I thought it was a pencil sketch, not a real study. They still amaze me. There are a few infections—HSV encephalitis and epidural abscesses as examples—where an MRI is the best test to make an early diagnosis.

In the future, with the help of the cardiac electrophysiologist, I will feel a little, very little, less worried about ordering an MRI with a pacemaker in place.

Rationalization

Nazarian, S. et al. A prospective evaluation of a protocol for magnetic resonance imaging of patients with implanted cardiac devices. Annals of internal medicine 155, 415-424 (2011).

Variations on a Theme

Pus is my life. It is what is on my car license plate. The word, not the material. PUS is also the acronym for the top three infections that lead to hospitalization: pneumonia, UTI, and soft tissue infection. I do not know if other specialties enjoy the variations on a theme that I do. Seems to me that all coronary artery disease or gastrointestinal bleeds are pretty much interchangeable, but that is probably due to my lack of insight into the subtleties of other medical disciplines. There is no end to the ways in which PUS, and pus, can manifest, including necrotizing soft tissue infections.

The patient is a diabetic who had had the end of a screwdriver accidentally (obviously) jammed into the end of her finger. Several days after the trauma she has an acute infection of the hand (it is red, hot, swollen, and tender). She also has blisters on the palm and necrosis at the end of the finger that suffered the penetrating trauma.

She is not septic nor does she have toxic or septic shock, and has the infection cleaned up in the OR, including removal of the distal part of the finger. No abscess or necrotizing infection was found anywhere but at the end of the finger. She initially gets vancomycin and then is changed to oral Bactrim (trimethoprim/sulfamethoxazole) and clinically relapses; the wound edges get necrotic and the erythema and blisters recur. So they call me. She is not systemically ill, so it's back to the OR where there are the same findings as before: lots of edema, no abscess, and no necrosis except at the wound edge.

The cultures you ask? *S. pyogenes* and MRSA.

Together they spell Meleney's synergistic gangrene. Part of the original description, from 1926, is available online, and worth a read if for no other reason than to see the medical writing style of the time.

Meleney's tends to be an indolent necrotizing infection of the abdominal wall soft tissues, although it can occur on the extremities. There are only twenty-one references on the PubMeds for

Meleney's, which kind of surprised me. Meleney's has not yet been described with MRSA, so you heard it here first.

Is this really a Meleney's? Microbiologically, yes. Clinically, maybe. The necrosis was slow, localized, and painful and the patient was not systemically ill. Against it, was the failure of antibiotics to arrest the infection, although it could be argued that the antibiotics were not optimal for the treatment, vancomycin being the archetype of lousy antibiotics and Bactrim not the best of our anti-Streptococcal antibiotics.

She slowly improved on penicillin and on vancomycin.

Rationalization

Brewer, G.E. & Meleney, F.L. Progressive Gangrenous Infection of the Skin and Subcutaneous Tissues, Following Operation for Acute Perforative Appendicitis: A Study in Symbiosis. Annals of surgery 84, 438-450 (1926).

Poll Results

My life is defined by

- pus. 15%
- debt. 12%
- Webster's. 3%
- love. 15%
- existential angst. 41%
- Other Answers 13%

    -more and more things that are smaller and smaller, until I vanish and all that remains is my smile.
    -nitric oxide.
    -my friends, family, and professional pursuits.
    -daily BS that is more and more annoying.
    -Monty Python as interpreted by Woody Allen and Groucho Marx
    -romanticism.

Beware the Inflection Point

MIDDLE-AGED. To me, the term does not mean the same today as it did 30 years ago. Middle-aged goes from 30 to 60 now; before it was 30 to 50; it will soon be 30 to 75. Middle age, from another view, is that point where the incidence of disease graphs start to slope up. Most disease incidence- versus-age graphs are flat in the teens and twenties, then have an inflection point where disease becomes more common. Maybe middle age should start at the inflection point, although that would lead to many definitions of middle age, depending on the disease. Safe to say, I am well past most inflection points, so am comfortably middle-aged. I do not want the end of middle age to be defined by, say, eligibility for an American Association of Retired Persons membership.

So by one definition, my patient is a middle-aged male who has three days of fevers and rigors and comes to the emergency room. He has had a heart murmur all his life (he was told it was the mitral valve), and now all his blood cultures are positive for gram-positive cocci in chains.

What does the echocardiogram show? A bicuspid aortic valve. This is a normal variation, sort of—about 1.3% of adults have only two leaflets of the aortic valve instead of three. But this variation is associated with narrowing of the valve (stenosis) and infection as people age and as the valve gets calcified. He is past the inflection point at which bicuspid aortic valves get infected or start to fail. More interestingly, he has a myocardial abscess. Usually streptococci do not cause an abscess this rapidly, so he must have had this infection longer than three days. I am even more convinced of this when the organism is identified as *Peptostreptococcus*, an anaerobic *Streptococcus*. That is not a common cause of endocarditis, a couple dozen reports at best, usually occurring on prosthetic valves.

It is probably the bicuspid valve that, for whatever reason, resulted in the abscess:

"Patients with BAV IE (n=50, 16%) were younger, had fewer comorbidities and a higher frequency of aortic perivalvular abscess (50%). Presence of BAV (OR 3.79 (1.97-7.28); p<0.001) was independently predictive of abscess formation. Early surgery was performed in 36 BAV patients (72%) with a peri-operative mortality of 8.3%, comparable to that of patients with tricuspid aortic valve IE (p=0.89). BAV was not independently predictive of in-hospital mortality (OR 0.89 (0.28-2.85); p=0.84) or 5-year survival (HR 0.71 (0.37-1.36); p=0.30). Age, comorbidities, heart failure, Staphylococcus aureus and uncontrolled infection were associated with increased 5-year mortality in BAV patients.
CONCLUSION:
BAV is frequent in adults with native aortic valve IE. Patients with BAV IE incur high risk of abscess formation and require early surgery in almost three-quarters of cases. IE is a severe complication in the setting of BAV and warrants prompt diagnosis and treatment."

Surgery found a large abscess that needed repair and valve replacement.

Therapy? As one reference notes, "very little is known about the appropriate therapy for these infections." It is usually penicillin, but my patient gets hives with penicillin, and there is now artificial material with the infections, making treatment more challenging. So for the short term, I am opting for overkill. Vancomycin and two weeks of gentamicin and metronidazole for its hopefully superior penetration into and around the valve ring, sutures, pledgets, and patch.

And he did just fine.

Rationalization

Choussat, R. et al. Perivalvular abscesses associated with endocarditis; clinical features and prognostic factors of overall survival in a series of 233 cases. Perivalvular Abscesses French Multicentre Study. European heart journal 20, 232-241 (1999).

Tribouilloy, C. et al. Clinical characteristics and outcome of infective endocarditis in adults with bicuspid aortic valves: a multicentre observational study. Heart 96, 1723-1729 (2010).

Capunitan, J.A. & Conte, H.A. Peptostreptococcus species: an unusual cause of infective endocarditis. Connecticut medicine 74, 93-96 (2010).

Cone, L.A., Battista, B.A. & Shaeffer, C.W., Jr. Endocarditis due to Peptostreptococcus anaerobius: case report and literature review of peptostreptococcal endocarditis. The Journal of heart valve disease 12, 411-413 (2003).

Poll Results

Middle age is defined by

- age. 10%
- attitude. 35%
- medication list. 23%
- when the years in front are fewer than the years behind. 27%
- my first divorce. 4%
- Other Answers 0%

Old Nodule

THE patient has had a stable two-centimeter lung nodule for several years and his doctors elected to watch it. He had no symptoms and no risks for cancer, but did eventually require an aortic valve replacement, and as long as they were in there, they resected the nodule.

Obviously not cancer, since I am writing about it, but old mold/yeast. Lots of old, degenerated fungal forms. Whatever this is, it has been there for a while. And they call me. What is it and what should we do about it?

I doubt it is an environmental mold like *Aspergillus*, which is not a common cause of a lung nodule that could be mistaken as a cancer. This is most likely an endemic fungus; the question is whether there is any suggestion it has disseminated—doubtful, since the resection should have cured the pulmonary infection.

History is not impressive: some travel to the Midwest (*Histoplasma* >> *Blastomyces*), scant travel to the Southwest (*Coccidioidomycosis*) so I bet *Histoplasma* >> *Cryptococcus*.

Wrong. Workup for all of the fungi is negative except for a low *Coccidioidomycosis* complement fixation. So it is probably old

Coccidioidomycosis. With a low complement fixation titer, I do not have to worry about disseminated disease, so nothing further to be done.

Although *Coccidioidomycosis* is found in the Southwest, the fungi can travel with the dust on trains, planes, and automobiles. We had a case here in the Northwest of *Coccidioidomycosis* pneumonia in someone who changed their girlfriend's car oil after the car, but not the patient, had returned from Sacramento.

I have been told anecdotes during my entire career of the travels of *Coccidioidomycosis* outside of endemic areas to cause disease. Although I suspect the stories of primary *Coccidioidomycosis* in non-travelers in Japan from airplane dust and in Boston from California oranges may be apocryphal, since I can't find them on the PubMeds. ID docs do love to tell their weird exposure histories.

I have seen two lobectomies and one pneumonectomy for fungal nodules. All were smokers; two had *Coccidioidomycosis*; one had *Cryptococcus*. All had a chronic cough, hilar adenopathy, atypical cells on bronchoscopy that were not diagnostic, and a chronic pneumonia that did not respond to antibiotics. Both of the *Coccidioidomycosis* cases were truckers who traveled between Los Angeles and Portland. None needed the pneumonectomy, but who besides an ID doc is going to think of an endemic fungus when the odds are it is cancer? I probably would not have considered *Coccidioidomycosis* either.

> *"The overwhelming majority of patients (93.3%) referred to "rule out" lung cancer were documented as having a neoplastic process, and only 1.3% had an infection. Fungal infections (histoplasmosis, cryptococcosis, coccidiomycosis) accounted for 46%, mycobacteria for 27%, bacteria for 22%, and parasitic lesions (dirofilariasis) for 5% of these infections."*

In medicine you have to play the odds, although oncologists, unlike those who treat infections, rarely treat without a pathologic diagnosis first, avoiding some disasters. Unfortunately, to make the diagnosis, sometimes you have to take a lesion out.

Rationalization

Chung, C.R. et al. Pulmonary coccidioidomycosis with peritoneal involvement mimicking lung cancer with peritoneal carcinomatosis. American journal of respiratory and critical care medicine 183, 135-136 (2011).

Petrini, B., Skold, C.M., Bronner, U. & Elmberger, G. Coccidioidomycosis mimicking lung cancer. Respiration; international review of thoracic diseases 70, 651-654 (2003).

Rolston, K.V., Rodriguez, S., Dholakia, N., Whimbey, E. & Raad, I. Pulmonary infections mimicking cancer: a retrospective, three-year review. Supportive care in cancer : official journal of the Multinational Association of Supportive Care in Cancer 5, 90-93 (1997).

Slow Infection

I DON'T remember my first kiss. I do remember my first great meal (Saint-Estèphe in Manhattan Beach), my first Bordeaux, my first profiteroles, and every good meal in my life.

I remember the best chicken and fries ever, in a bistro we wandered into to escape the rain after the Musée d'Orsay in France. That roast bird was one the many oral epiphanies of my life, never repeated. I like food, and, while less of a foodie than I used to be, eating is what I enjoy most. It is a miracle I don't weight 450 pounds. I also appreciate the whole slow food movement: local food cooked well and simply.

I also appreciate the slow infection. How is that for a segue? Kind of a stretch? Perhaps. Best I can do. You try writing case histories three times a week for four years and being clever.

The patient had rods (steel, not gram negative) put in her back three years ago and did fine in the interim. No symptoms whatsoever. Then an area over the rod came to head, a slightly red lump, with no other symptoms. Off the OR and pus was found tracking up and down the rods, which were removed. Gram stain was negative.

Three years post-op, no symptoms, and pus. A complete blood count, erythrocyte sedimentation rate (a measure of inflammation), and C-reactive protein (another measure of inflammation) are normal. Only one bug does that. *Propionibacterium acnes.* Coagulase-negative *Staphylococcus* is on the list as well, but usually doesn't fester without symptoms for years. Nor do fungi or other organisms, and I asked the lab to hold the cultures looking for *P. acnes*.

And what do you know, the cultures did indeed grow *P. acnes*.

My record is *P. acnes* presenting eight years after a craniotomy; evidently, that ain't nothin. One person presented twenty-three years post-op.

The keys to diagnosis? Hold the cultures:

"Culture time was long, on average 11.4 days."

P. acnes will be found as a cause of infections from craniotomies to hardware to line infections if you look for it. But you have to look for it, asking the lab to hold the cultures for two weeks.

The hardware is out. By the way: As I have said numerous times before, the three most dangerous words in medicine are: In. My. Experience. If anyone says ,"in my experience," in reference to choosing a therapy, they are talking out their burro. Or is it burrow? I guess I don't know an ass from a hole in the ground. Still, in my experience, patients respond to clindamycin better than to penicillin, but I work hard to ignore my experience and follow the literature, such as it is. I expect a cure with penicillin, which is the treatment of choice.

And it was so.

Rationalization

Levitt, M.R., Gabikian, P., Pottinger, P.S. & Silbergeld, D.L. Propionibacterium acnes osteomyelitis occurring 23 years after craniotomy: case report and review of literature. Neurosurgery 69, E773-779; discussion E779 (2011).

Lutz, M.F. et al. Arthroplastic and osteosynthetic infections due to Propionibacterium acnes: a retrospective study of 52 cases, 1995-2002. European journal of clinical microbiology & infectious diseases : official publication of the European Society of Clinical Microbiology 24, 739-744 (2005).

Martin-Rabadan, P. et al. Propionibacterium acnes is a common colonizer of intravascular catheters. The Journal of infection 56, 257-260 (2008).

Zeller, V. et al. Propionibacterium acnes: an agent of prosthetic joint infection and colonization. The Journal of infection 55, 119-124 (2007).

Poll Results

I most clearly remember, and my life was changed, by

- my first kiss. 10%
- my first patient. 10%
- my first day of internship. I have PTSD (post-traumatic stress disorder)[comment OK?]as a result. 21%
- my first great meal. 13%
- my first car. 31%
- Other Answers 15%

  -the birth of my children. (yes, that's number agreement.)
  -my first marriage (also my only so far— forty-three years and going)
  -the day my adopted daughter came home from China
  -my other daughter's birth

I Know It When I Se It

Some parts of my practice have declined over the past decade: Neupogen has led to fewer neutropenic fevers, highly active anti-retroviral therapy (HAART) has led to no AIDS-related opportunistic infections, and a variety of hospital initiatives have almost led to the extinction of hospital-acquired infections.

Fortunately, the bugs are becoming more resistant, so my work has not completely dried up. Resistance, said the Borg from *Star Trek*, is inevitable. Well, that is what the Borg would say if they had an interest in infection rather than in assimilation.

I am asked to see a patient with a urinary tract infection caused by what is billed as an MDRO, muti-drug resistant organism. She had been on multiple antibiotics in the past and has a variety of chronic medical problems, so I am expecting an extended spectrum beta-lactamase *E. coli* or mostly resistant *Pseudomonas*. But no. It is *Achromobacter xylosoxidans*. You may have heard of by its former name, *Alcaligenes xylosoxidans*. Before that, it was referred to as "Love Symbol #2." I have seen the bug on occasion in my career, almost always in diabetic foot infections, or so says my faulty memory. *Achromobacter xylosoxidans*is found in water and in rotting organic material, and the diabetic foot is, unfortunately, nothing if not rotting organic material. But the urine? The PubMeds help, as always:

"All patients had underlying diseases or urological abnormalities. The most frequent underlying diseases were solid or hematological malignancies (3 cases). Seven patients (77.7%) had urological abnormalities. Eight patients had symptoms of cystitis and 1 remained asymptomatic. Seven patients had community acquired UTIs. Clinical outcome was favorable in 5 patients after antibiotic treatment and recurrence occurred in 3 patients who had urological abnormalities. All isolates were susceptible to imipenem and piperacillin-tazobactam, 88.8% were susceptible to ceftazidime and 77.7% were susceptible to trimethoprim-sulfamethoxazole. High frequencies of resistance to ampicillin (100%), amoxicillin/clavulanic acid (78%), cefuroxime (100%), cefotaxime (67%), norfloxacin (89%), ciprofloxacin (78%), nitrofurantoin (89%) and gentamicin (67%) were observed."

Typical *Achromobacter*. Lots of antibiotic resistance. It is an MDRO, but is it an MDRO, one that warrants isolation? Vancomycin-resistant enterococci (VRE) and *MRSA* are easy to identify, but we keep coming up with difficulties in finding a simple way to identify gram-negative rods that require isolation.

To warrant isolation, the bug has to be resistant to the right collection of antibiotics AND be virulent, and how does one define virulent? MRSA is a worry; MRSE (*Staph epidermidis*) is not. Both are MDRO but only one is virulent. I also fret more when the resistance is likely on a plasmid, making the resistance mobile, rather that sitting on a chromosome, where it is perhaps more likely to stay put.

As Justice Potter Stewart, who helped tip the scales on many important decisions, said, "I shall not today attempt further to define the kinds of material I understand to be embraced within that shorthand description [MDRO]; and perhaps I could never succeed in intelligibly doing so. But I know it when I see it, and the *Achromobacter* involved in this case is not that."

Achromobacter may be MRDO, but it is not virulent, so it is not an MDRO. I decided the patient did not need MDRO isolation, just treatment for the UTI.

Rationalization

Tena, D., Gonzalez-Praetorius, A., Perez-Balsalobre, M., Sancho, O. & Bisquert, J. Urinary tract infection due to Achromobacter xylosoxidans: report of 9 cases. Scandinavian journal of infectious diseases 40, 84-87 (2008).

Love Symbol #2: https://www.pantone.com/about/press-releases/2017/the-prince-estate-and-pantone-unveil-love-symbol-number-2

Poll Results

I know it when I see it

- MDRO. 0%
- purulent material, I mean purient material;I mean just drop it, OK? 3%
- liberal bias . 8%
- conservative bias. 8%
- a most excellent infectious disease case. 79%
- Other Answers 3%

Forever

OVER three years ago I treated a patient with cryptococcal meningitis. It was *C. gattii,* which is now endemic in the Pacific Northwest. I think, but cannot prove, that he got it from bark dust. He works indoors in areas where there is a lot of bark dust and, well, dust. In the Northwest, *C. gattii* lives on fir trees and on other trees that are turned into mulch:

> *Tree types that were associated with high rates of positive swab and soil samples on the first sampling attempt included Douglas fir, alder, arbutus, red cedar, and Garry oak.*

In AIDS patients the serum cryptococcal antigen may not be of value:

> *We could not detect a significant correlation between sCRAG titer results of patients who had a clinical response to treatment and sCRAG titers in patients who experienced persistent disease, probable relapse, or definitive relapse of cryptococcal disease.*

With some amphotericin B (which was poorly tolerated) and then high-dose fluconazole, his meningitis and all of his symptoms were cured. He is back at work, no symptoms at all, normal spinal tap, and everything looks to be a cure. He is immunologically normal, at least as far as I can judge, so a cure should be assured. But.

The serum cryptococcal antigen is still positive. After 3.5 years it has slowly declined to 1:16. No focality on exam or review of systems that I can locate, but he has ongoing, but diminishing cryptococcal growth. He is in an odd equilibrium with the organisms: not ill, not cured.

The serum cryptococcal antigen (CRAG) titer is of more value, perhaps, in non-AIDS cryptococcal disease:

> *...serum CRAG titers among patients without HIV with CNS or pulmonary cryptococcosis declined during treatment and no relapse was noted when serum CRAG titers were*

But who would want to stop therapy for an infection when you could still measure body parts in the blood? Not me. One of the issues that keeps showing up in infections with increased frequency are polymorphisms in toll-like receptors that lead to the inability to control a variety of infections. I could not find a report, yet, in the PubMeds in humans, but it is only a matter of time. In mice, it is the TLR9 that is important for the control of *Cryptococcus*. Or perhaps his extended disease is *C. gattii* being less responsive to azoles.

Maybe it is the host; maybe it is the yeast, maybe a bit of both. I guess for now, push ahead. He is slowing getting better, but someday, when his antigen is negative, I will have to consider stopping the fluconazole. Or perhaps, unlike love, *Cryptococcus* is forever.

Rationalization

Kidd, S.E. et al. Characterization of environmental sources of the human and animal pathogen Cryptococcus gattii in British Columbia, Canada, and the Pacific Northwest of the United States. Applied and environmental microbiology 73, 1433-1443 (2007).

Aberg, J.A., Watson, J., Segal, M. & Chang, L.W. Clinical utility of monitoring serum cryptococcal antigen (sCRAG) titers in patients with AIDS-related cryptococcal disease. HIV clinical trials 1, 1-6 (2000).

Lin, T.Y. et al. Cryptococcal disease in patients with or without human immunodeficiency virus: clinical presentation and monitoring of serum cryptococcal antigen titers. Journal of microbiology, immunology, and infection = Wei mian yu gan ran za zhi 42, 220-226 (2009).

Wang, J.P., Lee, C.K., Akalin, A., Finberg, R.W. & Levitz, S.M. Contributions of the MyD88-dependent receptors IL-18R, IL-1R, and TLR9 to host defenses following pulmonary challenge with Cryptococcus neoformans. PloS one 6, e26232 (2011).

Poll Results

Forever describes

- *Twilight* movies. It sure seems that way. 7%
- the last trimester. 12%
- nothing except the earth and sky. Damn I hate Kansas. 23%
- Christmas Eve for a six- year- old. 26%
- love, you cynical old coot. 23%
- Other Answers 9%

Overcoming the I/O Bottleneck to the Faulty RAM

Sepsis is not an uncommon reason for consultation. The patient is a middle-aged female who presents with three days of fevers and progressive abdominal pain. The workup is negative for a cause for the sepsis, and she rapidly slides into multi-organ system failure despite antibiotics and the usual ICU care.

They call me when the blood cultures grow *S. pyogenes*: Group A *Streptococcus*.

Group A streptococcal infections usually have a soft tissue source. Patients are septic with either cellulitis or a necrotizing infection. But in this case, a physical exam and a pan-scan yield nothing but marked grimacing with rebound and guarding with abdominal palpation, a sign of an intra-abdominal infection.

The exam also has the suggestion of toxic, in addition to septic shock. The palms and chest have a sunburn erythroderma and the calcium level is low. The eyes, soles, and tongue, however, are normal. A couple of years ago I wrote about a case of cellulitis with toxic shock syndrome (TSS), but the classic skin and mucocutaneous findings occurred later, after the patient was off high-dose pressors. He eventually desquamated, confirming that it was a TSS rash. So I wonder if hypoperfusion of the skin will mask the skin findings of TSS.

There were enough TSS findings that I recommended treating for TSS with penicillin, clindamycin, and intravenous immunoglobulin. But what is the source? Group A TSS has a source of the infection. Gynecological? Nothing on abdominal CT and no recent gyn procedures. Just abdominal pain and a negative CT.

I have been been doing ID for almost thirty years. I have easily seen 20,000 consults, three times that many curbsides walking the halls and ten times that many phone calls. My clinical experience is huge, but my ability to recall that experience is increasingly flawed. The I&O to the hard drive of my memory is an increasing bottleneck and the sectors of my RAM are increasingly suspect. But the nice thing about practicing in the twenty-first century is that I don't have to remember things—I just I have to remember that I remember something and then search PubMed.

I long ago gave up on the idea that I, the conscious me, knows what I am doing. Whatever part of me analyzes information and spits out a diagnosis is not under conscious control. Much of my discussion with residents on how I reach a diagnosis is a post hoc ergo propter hoc rationalization of a process over which I no longer have control. There was a time when I ran down lists and differential diagnoses, and I still do, but much of that has been internalized and below rational, conscious thought. It is still spooky how ideas burble up from somewhere into conscious awareness. I am just the ghost in the machine. Or crazy.

So initially I was flummoxed. No source. TSS/septic shock. Abdominal pain. I agreed to a transesophageal echocardiogram with little enthusiasm, given the rarity of Group A streptococcal endocarditis, and suggested a formal gyn evaluation. I wrote my note and then started to dictate. The dictation was distracted, slightly unfocused, as I had this niggling idea I was missing something I had seen before—some vague idea that it was in front of me.

As I was getting to the discussion part of the dictation I paused, opened PubMed, and entered, "spontaneous bacterial peritonitis" (SBP) and "pyogenes" as search terms into the PubMeds. Bingo.

Group A *Streptococcus* causes SBP with septic and toxic shock, often in normal women without ascites. Later they tapped the scant fluid in the abdomen and it was cloudy with gram postitive cocci in chains. Diagnosis confirmed.

I maybe remember I think perhaps I had heard of or seen a similar case once upon a time. Or did I? I do not know what led me to that search in the middle of a dictation. It is nice as I get older that if I remember that I might remember something, I have PubMed to back me up. It doesn't change the treatment, but at least we have an explanation of how, if not why, the patient is ill.

Rationalization

Monneuse, O. et al. Diagnosis and treatment of spontaneous group A streptococcal peritonitis. The British journal of surgery 97, 104-108 (2010).

van Lelyveld-Haas, L.E., Dekkers, A.J., Postma, B. & Tjan, D.H. An unusual cause of a spontaneous bacterial peritonitis in a young healthy woman. The New Zealand medical journal 121, 82-85 (2008).

Brase, R., Kuckelt, W., Manhold, C. & Bohmert, F. [Spontaneous bacterial peritonitis without ascites]. Anasthesiologie, Intensivmedizin, Notfallmedizin, Schmerztherapie : AINS 27, 325-327 (1992).

Poll Results

I am

- conscious. 3%
- an automaton with the illusion of consciousness. 28%
- a ghost in the machine. 34%
- the dream of a butterfly. 9%
- in the matrix and will NOT take the red pill. 25%
- Other Answers 0%

Grumpy Old Man on Call

WEEKENDS can be busy, especially Saturday when I have to get to know my colleagues' patients. I share calls with two other ID docs, and weekends mean responsibility for seven hospitals over a wide geographic range. It can be done thanks to hospitalists, residents, and electronic medical records (EMRs). One good thing about the EMR is that I no longer have to be onsite to review labs and studies. But I have a lot of territory to cover and many ill patients to see, and there are the new consults.

I get consults for a variety of reasons and I am always happy to be called. I like seeing consults. From my perspective, there is almost no such thing as a bad consult. Except maybe. I hate second opinions generated not by patients or by attendings, but by dissatisfied family members. I always think of the old joke:

Doctor: I think you are crazy.

Patient: I want a second opinion.

Doctor: OK. You are ugly, too.

So when I get the request to see an elderly man with severe chronic obstructive pulmonary disease (COPD) who keeps being admitted for exacerbation because the daughter wants an ID consult before they will take him home, I bitch and moan and whine. It is obvious from the discussion with the referring doc that my services are not needed. I have ill patients to see all over the city and a hand-holding consult is the last thing I want to do. But I can't say no. So I make the drive to the outlying hospital, review the chart, and, always remembering it is my job to take care of everyone to the best of my ability, leave the grumpy old man persona at the nursing station and see the patient. You always have to leave your baggage outside the patient's room; being professional is nothing if not being able to fake it well. I am talking medicine here, you know.

Within the hour it is obvious that the patient has no infection that requires treatment, that it is all severe, progressive COPD and right heart failure. A bronchoscopy and a CT scan are unimpressive for pneumonia or for other processes, and the cultures are colonization. I tell the patient the whys and wherefores of her process and she seems satisfied. No infection here. Keep on moving.

But that doesn't mean there isn't something to be learned. I tell the residents it is the great docs who make the great cases, and every patient is an opportunity to ponder the cause and treatment of disease, especially since all diseases, at least all interesting diseases, are infectious.

The sputum cultures taken when she was an outpatient have grown *Pseudomonas* and *Stenotrophomonas*. It is not a cause of infection clinically or radiographically, but it is an interesting question what role colonization with bacteria plays in COPD exacerbations and subsequent morbidity in some patients. Well defined in cystic fibrosis progression, *Pseudomonas* has bad associations in COPD as well:

> *PA isolation in sputum in patients hospitalized for acute exacerbation of COPD is a prognostic marker of 3-year mortality.*

and

> *"These results provide the first evidence supporting the hypothesis that P. aeruginosa causes chronic infections in COPD, with patterns of infection and evolution that resemble those observed in cystic fibrosis."*

There are no good treatment trials, at best a series of anecdotes, and the plural of anecdote is anecdotes, not data.

Aerosol administration of high-dose tobramycin in non-CF bronchiectatic patients for endobronchial infection with PA appears to be safe and decreases the risk of hospitalization and PA density in sputum. Nevertheless, pulmonary function and quality of life are not improved, and the risk of bronchospasm is appreciable.

In the end, I lateraled back to the consulting physician. For me it was an opportunity for book learning; I don't take care of COPD exacerbations, and as the weekend doc at another physicians' hospital, I deferred to others as to the utility or not treatment.

But never pass up an opportunity to Google a case.

Rationalization

Almagro, P. et al. Pseudomonas aeruginosa and mortality after hospital admission for chronic obstructive pulmonary disease. Respiration; international review of thoracic diseases 84, 36-43 (2012).

Murphy, T.F. Pseudomonas aeruginosa in adults with chronic obstructive pulmonary disease. Current opinion in pulmonary medicine 15, 138-142 (2009).

Dal Negro, R., Micheletto, C., Tognella, S., Visconti, M. & Turati, C. Tobramycin Nebulizer Solution in severe COPD patients colonized with Pseudomonas aeruginosa: effects on bronchial inflammation. Advances in therapy 25, 1019-1030 (2008).

Drobnic, M.E., Sune, P., Montoro, J.B., Ferrer, A. & Orriols, R. Inhaled tobramycin in non-cystic fibrosis patients with bronchiectasis and chronic bronchial infection with Pseudomonas aeruginosa. The Annals of pharmacotherapy 39, 39-44 (2005).

Martinez-Solano, L., Macia, M.D., Fajardo, A., Oliver, A. & Martinez, J.L. Chronic Pseudomonas aeruginosa infection in chronic obstructive pulmonary disease. Clinical infectious diseases : an official publication of the Infectious Diseases Society of America 47, 1526-1533 (2008).

Poll Results

I get grumpy with calls concerning

- refill of pain meds on the weekend. 12%
- second-opinion consults. 12%
- VIP consults. 16%
- nursing homes where the calling person has a question but can answer none of yours. 36%
- that one doc who calls all the time but has never actually given me a paying consult. Ever. 12%
- Other Answers 12%

Sinking Feeling

I GET a call from the ER. "Do you remember patient Hungadunga?"

Not offhand. I barely remember yesterday anymore. I call my wife "hun" and "sweety" since I have trouble recalling her name.

"Liver failure. You treated him for four weeks for endocarditis. Stopped the antibiotics about a week ago."

Oh yeah. Hungadunga. *Streptococcus mitis* in the blood in multiple blood cultures. Now I remember.

He is back in with fevers and abdominal pain.

Cue the sinking feeling. ID is a practice where most of the time, when you expect to cure a patient, you do cure the patient. Especially a susceptible streptococcal endocarditis, so when you get a call that the patient is back, you think—or at least I think—"Did I screw up? Did I miss something? Should I have stopped the antibiotics?"

"Well," I said, "maybe it is a relapse; maybe it is a new infection given his end-stage liver disesase, so kill everything and we will see what grows in the blood cultures."

But it is night, at home. I can't remember making any potential errors, but falling back asleep is not so easy. I perseverate on the issue far longer than I should. My dad said after he retired that the most tiring part of medicine was worrying, not the hours, and at times I appreciate the sentiment. He was a cardiologist, so he had a lot more to worry about than an ID doc.

Curiously, I thought I could find the answer to anything on the interwebs, but my Google-fu must be weak, as I can't find what the physiology behind that the awful feeling in the pit of the stomach. The closest is a patient who had the feeling as an aura preceding his seizures, but the literature is lacking in a good explanation for the feeling. There is a research project for some-one.

Fortunately for my propensity for anxiety over bad outcomes, a review of the chart the next day does not reveal any mistake I could see, so I am somewhat relieved . . .until the blood cultures start growing Streptococci.

Crap. The sucking in the pit of the stomach returns. They are gram-positive cocci in pairs only, not in chains. Not like a *S. mitis*.

And the next day? Whew. *Streptococcus pneumoniae* in the blood. And, thanks to vancomycin and cefepime, the patient is doing well. Double whew. Sensitivity tests are not done yet, but given the four weeks of ceftriaxone, it will be interesting to see what the minimum inhibitory concentration to penicillin is. I hope I have not helped breed a resistant *Pneumococcus*.

There are a lot of reasons why the patient could be bacteremic, given his liver disease. It would be fun to check both the serotype of the *S. pneumoniae* and to see if, because of advanced cirrho-sis, he has a mannose-binding lectin deficiency. But it will not help the patient and there is no treatment, so why bother? The current buzzword, not quite applicable in this context, is "mean-ingful use." Labs have to be justified in making a difference in the patient's care. The good, bad, or indifferent old days, when tests could be ordered to maximize diagnostic certainty, or just to scratch a curious itch, are going to be an intellectual exercise of the past. Oh well. Good thing I am near the end, rather than at the beginning, of my career.

Postscript

The MIC to penicillin was low (the bacteria were not resistant), so I suspect he acquired the *Pneumococcus* after he finished his antibiotics.

Rationalization

Valles, X. et al. Serotype-specific pneumococcal disease may be influenced by mannose-binding lectin deficiency. The European respiratory journal 36, 856-863 (2010).

Manford, M. & Shorvon, S.D. Prolonged sensory or visceral symptoms: an under-diagnosed form of non-convulsive focal (simple partial) status epilepticus. Journal of neurology, neurosurgery, and psychiatry 55, 714-716 (1992).

Poll Results

When my patients unexpectedly have a problem I

- count the tiles in the ceiling long into the night. 20%
- blame someone else. 10%
- deny there is a problem. 7%
- get that horrible sinking feeling and take a proton pump inhibitor 40%
- thanks to Zoloft and Ativan, I see no problem here . 23%
- Other Answers 0%

Shingles

I HAD shingles when I was young. It was V1 distribution when I was an ID fellow. It involved my eye and is part of the reason why I no longer wear contacts. I still remember seeing the ophthalmologist and his mentioning that he understood that shingles were somehow related to chicken pox. I was not comforted.

If you live long enough, you will get shingles, and you should hopefully get it only once in a lifetime. I have been boosted at least twice in my life: the shingles episode, and then I was exposed when my oldest child had chicken pox before there was a vaccine. I am still going to get the vaccine in a couple of years, because better safe than sorry. I had no postherpetic neuralgia, but the older you are, the worse the shingles, the more likely there will be postherpetic neuralgia, and I do not want a recurrence.

The patient is in his eighties, has thoracic shingles, and is admitted for altered mental status. An MRI is negative, but he has white blood cells and elevated protein on the spinal tap. A quick exam of the skin shows a few scattered chicken pox lesions on the extremities and on the abdomen. Neither of these is unusual. A few disseminated lesions are common with shingles, as is a spinal tap consistent with mild viral meningitis. VZV is also not an uncommon cause of aseptic meningitis without shingles or chicken pox:

VZV DNA was detected in 2 of 45 samples (4.4%).

Although one has to wonder if it is more like a fever blister: the virus reactivates in the cerebrospinal fluid because of another CNS process. Causality with VZV is sometimes suspect.

What is more important than the spinal tap is the clinical syndrome. In this case, the patient remained altered for days despite intravenous acyclovir. It was encephalitis rather than meningitis—a more unusual manifestation of VZV. VZV encephalitis is often due to a CNS vasculitis, although it is not typical in this patient, as the granulomatous arteritis of shingles and the small-vessel encephalitis tend to happen late in the clinical course and in the immunocompetent. Also against VZV vasculopathy is the lack of MRI changes, but with the clinical shingles, there appears to be no other reason for the altered mental status.

Whether it was VZV vasculitis/encephalitis or meningoencephalitis, I will never know. I do not know if the patient had the zoster vaccine, nor if it would have made any difference. Still, I hold to the idea the three greatest inventions for human health have been vaccines, flush toilets, and antibiotics. I make use at least two whenever the opportunity presents itself. Or is that oversharing?

Rationalization

Persson, A., Bergstrom, T., Lindh, M., Namvar, L. & Studahl, M. Varicella-zoster virus CNS disease--viral load, clinical manifestations and sequels. Journal of clinical virology : the official publication of the Pan American Society for Clinical Virology 46, 249-253 (2009).

Franzen-Rohl, E., Tiveljung-Lindell, A., Grillner, L. & Aurelius, E. Increased detection rate in diagnosis of herpes simplex virus type 2 meningitis by real-time PCR using cerebrospinal fluid samples. Journal of clinical microbiology 45, 2516-2520 (2007).

Gilden, D.H., Kleinschmidt-DeMasters, B.K., LaGuardia, J.J., Mahalingam, R. & Cohrs, R.J. Neurologic complications of the reactivation of varicella-zoster virus. The New England journal of medicine 342, 635-645 (2000).

Gilden, D., Cohrs, R.J., Mahalingam, R. & Nagel, M.A. Varicella zoster virus vasculopathies: diverse clinical manifestations, laboratory features, pathogenesis, and treatment. The Lancet. Neurology 8, 731-740 (2009).

Poll Results

The greatest invention in medicine is

- the electronic medical record. 2%
- vaccines. 57%
- antibiotics. 34%
- health insurance. 1%
- managed care. 0%
- Other Answers 6%
    - -soap and clean water.
    - -bogs.

Contaminants?

Positive blood cultures come in three flavors.

1) Pathologic. It is real and a manifestation of a disease.

2) Contaminant. Usually a coagulase-negative *Staphylococcus* or a Diphtheroid. Can be safely ignored—most of the time.

3) In between. Sometimes the blood cultures are real, as the bug really is in the blood, but not it is not causing disease. Bacteremia is common, with very low numbers of organisms, and on occasion the blood cultures will catch an alpha *Streptococcus* from chewing gum or some such. Or someone who is found down will grow skin *Staphylococcus* in their blood. In both instances, the bacteria in the blood are probably real and clinically unimportant. They can usually be safely ignored.

Whether or not to ignore blood cultures depends on the clinical presentation and risk factors; as a rule, no organism in a blood culture can be ignored out of hand.

Also, some organisms always have to be treated. *S. aureus* and *E. coli* are common organisms that you always have to presume are the real deal. About once a year I get a consult for a patient who has a high fever, rigors, and has *E. coli* in the blood, but evaluation for a source yields zip. My response is to give my best Gallic shrug. *C'est la vie.* Clinically real, probably transient, of no long-term consequence, and I always give a course of antibiotics.

S. aureus, the emperor of pathogens, is always always always the real deal. Well, almost always.

The patient is elderly and admitted with a stroke. Smoker, hypertension, increased lipids, doesn't take his meds or see his doctor. He is admitted to the ICU, is afebrile, and has a normal white blood cell count. For reasons I cannot glean from the electronic medical record (as if I can gather any narrative from EPIC), one set of blood cultures are drawn and one bottle of the set grows *S. aureus* and coagulase-negative staphylococcus. They call me. There are no stigmata of endocarditis or other infection—except for the stroke, which could be an endocarditis embolic surrogate. For the last two days he has had no signs or symptoms of infection. Repeat blood cultures are negative, albeit after a slug of vancomycin.

I have said this before, but the world would be a much better place if, upon awakening, everyone would ask themselves, how can I make Mark Crislip's life better? (And by extension, improve the lives of infectious disease doctors everywhere?) The answer is quite simple. Before giving antibiotics for positive blood cultures, REPEAT THE BLOOD CULTURES AAAARRRGGGHHHHHHH. Sorry. I get emotional on the topic.

But. He has horrible psoriasis—oozing, weeping lesions everywhere, including next to the antecubital fossa of the elbow: right where they drew the blood cultures. To have a severe skin disease is to be colonized with *S. aureus*. For psoriasis:

> *"There was a statistical difference in cultivation of S. aureus between lesional (64%) and non-lesional skin (14%) in patients with psoriasis (p = 0.037). S. aureus was cultivated from the nares in 25 (50%) of 50 patients with psoriasis and in 17 (34%) of 50 healthy controls (p > 0.05)."*

And for other skin diseases,

> *"Bacteria were isolated from 70.2% of lesional and 32.7% of nonlesional skin samples from patients with eczema, of which S. aureus accounted for 47.3% and 27.9%, respectively. Bacteria were isolated from 74.8% of lesional and 34.5% of nonlesional skin samples from patients with AD (atopic dermatitis), of which S. aureus accounted for 79.8% and 80.5%, respectively."*

For *S. aureus*, oozing skin must be like an all-you-can-eat buffet. Mmmmm. Oozing skin.

For one of the rare times in my career, I actually suggested a transesophageal echocardiogram. While I was clinically convinced that the *S. aureus* was a contaminant, the stroke gave me pause. The TEE was pristine. For the first time since my fellowship, and in a similar scenario, I declared a *S. aureus* in the blood OK to ignore.

I have the power, and I am not afraid to use it.

Rationalization

Balci, D.D. et al. High prevalence of Staphylococcus aureus cultivation and superantigen production in patients with psoriasis. European journal of dermatology : EJD 19, 238-242 (2009).

Gong, J.Q. et al. Skin colonization by Staphylococcus aureus in patients with eczema and atopic dermatitis and relevant combined topical therapy: a double-blind multicentre randomized controlled trial. The British journal of dermatology 155, 680-687 (2006).

Poll Results

To make Mark Crislip's life better I will

- repeat blood cultures before antibiotics. 21%
- avoid any future discussion of iron. 19%
- buy his books. 3%
- ignore spelling and grammar in his entries and focus on the content. 38%
- laugh even if his jokes aren't funny. 16%
- Other Answers 3%

  -all of these except #1. I am a microbiologist, not a physician. I can't order blood cultures or antibiotics.

Water You Talking About

"Son, stocks may rise and fall, utilities and transportation systems may collapse. People are no damn good, but the cultures will always tell the truth." — Lex Luthor. Sort of.

I HAD this patient who kept having fevers and positive blood cultures. She was in the hospital for acute onset fevers, delirium, and hypotension. She had an issue with chronic opiate use due to chronic pain from an infected hip, but denied using IV drugs.

The first time was a *Sphingomonas paucimobilis.*

Two weeks later it was *Stenotrophomonas maltophilia.*

After an extensive workup with no reason found for the bacteremias, I told the patient that the only way that these organisms could get into the bloodstream, was if the water was injected into the vein. Are you doing that?

She looked me right in the eyes and solemnly swore no way, not me, didn't happen. "I can't explain how those bacteria got in my blood, but I didn't do it."

Generally speaking, patients tell you the truth when they are hospitalized, since deliberately giving the wrong information to your physician could lead to death. I would rather take care of a heroin user than a bank president. Drug users are usually upfront about their habits. I was skeptical, but what can you do? I do not like to blame the patient for their disease, since if I am wrong, I could potentially miss an important diagnosis.

Within hours of the discussion she had a new set of fevers, hypotension, and delirium and the blood grew *Stenotrophomonas maltophilia* again. Bingo. I tested the water in her bathroom, feeling confident I would grow the *Stenotrophomonas* and have a smoking gun, although a gun can't smoke much in water. No luck! The environmental cultures were negative.

Years ago I had a heroin user who would come in every month or two with *Acinetobacter* sepsis with a negative workup. She swore she used clean needles when she injected her heroin. I eventually discovered she used water either from the lobby of a hostel or from the women's room at a local fast-food restaurant. I should have cultured both, but I told her that one of those was probably the source of her bacteremia, and when she stopped using that water, she quit coming into the hospital septic. Or maybe her habit killed her. I don't know. But the cultures told me the probable source of the infection. Her bugs HAD to come from water, injected directly — no other explanation.

Fast-forward two months. She is readmitted with a septic hip. It grows *C. albicans*. While in the hospital she has fevers and her blood grows a *C. parapsilosis*. Again, environmental organisms. Again, denial.

And then, as an outpatient, her primary has a sit-down confrontation, and she admits that yes, she is dissolving her pain meds in tap water and injecting them. She didn't use hospital water, but had brought water from home, hence the negative cultures. The water was the source. The *Candida*? She also had a habit of pretending to swallow pain pills, hiding them under her tongue next to the antibiotic-induced thrush, to inject them later.

The cultures never lie.

Rationalization

Perola, O. et al. Recurrent Sphingomonas paucimobilis -bacteraemia associated with a multi-bacterial water-borne epidemic among neutropenic patients. The Journal of hospital infection 50, 196-201 (2002).

Adjide, C.C. et al. [Stenotrophomonas maltophilia and Pseudomonas aeruginosa water-associated microbiologic risk assessment in Amiens' University Hospital Centre]. Pathologie-biologie 58, e1-5 (2010).

Poll Results

I least want to take care of

- IV drug abusers. 5%
- CEOs. 14%
- patients who bring in a folder full of University of Google printouts. 47%
- health care workers. 0%
- the family of health care workers. 30%
- Other Answers 5%

    -chronic pain + borderline personalities.

....

FUD

I USUALLY like to write about the cases I have recently seen, but last week I was dead in the water. Not a single consult at any of my three hospitals. I have had slow weeks before, but usually in August, not in midwinter. It is probably a combination of a slow influenza season combined with my hospitals having almost completely eradicated hospital-acquired infections. And with my help. Nothing like putting yourself out of business.

So instead of discussing a recent fascinoma, I am going to switch to grumpy old man mode and complain. I have an AARP card, so I have the power and am not afraid to use it. And get off my lawn.

I do not know how you think of the world, but I think of it as covered with microorganisms. In some places it is thicker, and in others it is thinner, but everything is covered with bugs. If you had Superman's vision and germophobia, you would have to fly into the sun to escape the bacteria. You would never touch anything, much less your spouse and kids.

I do not worry about most bacteria and do not worry overmuch about infections. Only a small fraction of organisms are potentially pathogenic, and even those leave us alone most of the time. Hence, the need to work for food. Everyone knows the five-second rule about food. I have the five-day rule. As long as ants are not on it, I'll pick it up and eat it.

We wallow in bacteria acquired from each other and from the environment every second of every day. It is probably to our benefit, and I have always been sympathetic to the hygiene hypothesis. Yet once a study gets published about the ubiquity of bacteria, you can be confident someone will manufacture Fear, Uncertainty, and Doubt.

Some papers have demonstrated that flushing a toilet without a lid aerosolizes *Clostridium difficile* and other pathogens into the environment. That is useful information for infection control in the hospital and for households where a member has gastroenteritis, although I suspect given the close contact of families, worrying about the toilet flush will be too little, too late. And familial spread of *C. difficile* is rare.

So you have a world covered in bacteria and the potential to spread bacteria with a toilet flush. Maybe you should worry if a family member has *Shigella*, but otherwise, your family's flush poses an infinitesimal risk to your health. So when I read the following in my paper:

> *"A recent study showed that flushing a toilet with the lid up can spray millions of bacteria like E. coli into the air, which can land on your toothbrush. So, it's best to put the lid down before each flush. But if you or others in your household are erratic about this, take care to keep your brush healthy and clean by storing it upright inside your medicine cabinet, not on the bathroom counter top. You can also disinfect your toothbrush periodically by microwaving it on high for 15 seconds."*

I think, "How stupid." Unless the toothbrush was being used to scrub the toilet, the chance that there would be an infectious inoculum of even *Shigella* would be homeopathically minuscule. Folks, there are real risks in the world. If you are anxious about *E. coli,* autoclave your dog. Your toothbrush? Not so much.

Rationalization

Q&A: What is a pathogen? A question that begs the point. http://www.biomedcentral.com/1741-7007/10/6

Barker, J. & Jones, M.V. The potential spread of infection caused by aerosol contamination of surfaces after flushing a domestic toilet. Journal of applied microbiology 99, 339-347 (2005).

Best, E.L., Sandoe, J.A. & Wilcox, M.H. Potential for aerosolization of Clostridium difficile after flushing toilets: the role of toilet lids in reducing environmental contamination risk. The Journal of hospital infection 80, 1-5 (2012).

Gerba, C.P., Wallis, C. & Melnick, J.L. Microbiological hazards of household toilets: droplet production and the fate of residual organisms. Applied microbiology 30, 229-237 (1975).

Pepin, J., Gonzales, M. & Valiquette, L. Risk of secondary cases of Clostridium difficile infection among household contacts of index cases. The Journal of infection 64, 387-390 (2012).

Johnson, J.R. et al. Phylogenetic and pathotypic similarities between Escherichia coli isolates from urinary tract infections in dogs and extraintestinal infections in humans. The Journal of infectious diseases 183, 897-906 (2001).

Poll Results

When it comes to bugs, I

- never worry about bacteria in my environment. 13%
- never touch anything without ethanol foam first. 3%
- wear a total body condom like Frand Drebin 24 x 7. 8%
- am convinced that there are no germs. It is a myth generated by Big Pharma to sell drugs. 15%
- am only concerned in the hospital and in the clinic. And that guy coughing continuously in the airplane one row back. 58%
- Other Answers 2%

Doing the Impossible

Sometimes I am asked to do the impossible. Go faster than light? Nope. Divide by zero? Hardly. Have equitable universal health care? Ho ho ho.

It is impossible to prove a negative, or so I have been told. You can't prove that there is no Tooth Fairy or that there are effects from homeopathy, just that the preponderance of evidence suggests neither exists in reality as we understand it. But if The Rock shows up in my bedroom with a pair of wings trying to exchange an incisor for a quarter, I will have to change my mind, or perhaps I am hallucinating after an overdose of Oscillococcinum.

Today I was asked to "clear" a patient for open heart surgery and declare that his fevers were not from an infection that will put his soon-to-be-new aortic valve at risk. That is, of course, an impossible consult.

The patient had no infection symptoms prior to his admission: nothing on review of systems, just living *la vida loca.* Then he has new chest pain, gesturing over his heart with a squeezing motion. I haven't seen a patient do that for years, MI aka heart attack being rare in my world. With the chest pain are fevers and a rigor. It occurs a second time (pain, then fever/rigor), and this time he is admitted to the hospital and is found to have an MI and tight aortic stenosis. But no infection.

He has two more fevers and rigors, but no pain, and then nothing.

Is it OK to operate? I guess. I can't find a reason for the fevers and rigors. Certainly, acute infections of all kinds increase the risk of vascular events: strokes, heart attracts, and pulmonary embolism. Infections are proinflammatory, and inflammation is prothrombotic, and thrombosis leads to vascular events. So did the infection come first? Not that I can glean from my history and physical, labs, and x-rays.

I think his fevers are from the MI. You have to go back to the days when there was almost nothing to do for MIs, 1974, for the last review:

> *"The seven-day point prevalence for fever was 65.3 per cent, with a maximal incidence of 72 per cent on the second day. No significant correlation could be found between maximal temperature levels, degree of enzyme elevation, and leukocyte count."*

And a fever from MI should last no more than four days, and his resolved in three. Of course, if the blood cultures pop positive then I am wrong. It only takes one counterexample to disprove a hypothesis. But for now, like declaring that all crows are black, I declare the patient uninfected. Just ignore the ever so slight quiver of uncertainty in my voice.

P.S. The blood cultures were negative.

Rationalization

Gibson, T.C. The significance of fever in acute myocardial infarction: a reappraisal. American heart journal 87, 439-444 (1974).

Poll Results

Impossible is

- oscillococcinum having effects. 19%
- equitable universal health care. 5%
- dividing by zero. 8%
- a chapter with no spelling or grammar errors. 19%
- a medical field more endlessly interesting than infectious diseases. 46%
- Other Answers 3%

  -keeping track of mutated bugs.

The Oldest Diagnostic Uncertainty

THE patient is here to see me for a positive QuantiFERON gold, the newfangled TB test. She needs immunosuppressive therapy for another condition, and so I am consulted to make sure she doesn't have TB before they knock down her immune system. I see these now and then.

The underlying disease is curious. She is an older white female (hence the pronoun) who has had an extensive workup in three hospital systems over a decade (Kaiser, University and now mine) with the final thought that she has sarcoid. The first two workups were close, but never had a definitive biopsy, as if noncaseating granulomas were diagnostic.

It is an odd case of sarcoid: no adenopathy and unilateral progressive infiltrates with small abscesses and scarring. The upper half of the right lung looks like a chronic infection. But too-numerous-to-count numbers of cultures have never demonstrated a pathogen, so it seems to be sarcoid. The patient has no symptoms of consumption; in fact, she has gained weight, so a chronic infection seems unlikely. Her pulmonary function has been declining, and she is due for a course of steroids, hence the evaluation for latent TB.

She has zero risk factors for TB. A middle-class Oregonian with no travel or exposures even barely suggestive of TB, so is this a false positive? No way to know, although false positives can occur with other mycobacteria such as *Mycobacterium kansasii*, *Mycobacterium szulgai*, and *Mycobacterium marinum*.

> *"The U.S. born population with a low prevalence of latent MTB are more likely to have false-positive QFT-G tests, albeit it can occur up to 9% of the time in foreign-born subjects in the U.S. Therefore, the use of QFT-G alone for the detection of latent MTB in healthy populations could be problematic."*

The TB skin test is negative, but that doesn't help with the clinical decision as sarcoid leads to anergy:

> *"We studied 38 patients with sarcoidosis (22 men, 16 women; mean age 42.5 years), 30 patients of TB (18 pulmonary, 12 extrapulmonary) and 30 healthy controls. Patients with sarcoidosis were more likely to have a negative TST compared to healthy controls (89.5% vs. 60%, p = 0.004) or TB (89.5% vs. 23.3%, p < 0.001). However, QFT positivity was not significantly different in sarcoidosis compared to controls (34.2% vs. 50%, p = 0.19), but was higher in TB patients as compared to sarcoidosis (60% vs. 34.2%, p = 0.03)."*

If the positive QuantiFERON is the real deal and she gets steroids, the TB, if really there, could go bananas. Could it be due to the sarcoid? Unlikely. Although my bias is any disease that causes granulomas is due to an infection, sarcoid does not appear to increase the rates of a positive QuantiFERON gold:

"A total of 44 sarcoidosis patients (22 men) with a median age of 39 y (range 25–59 y) were enrolled; 93% had a negative QFN test result and 7% had an indeterminate result. Forty-three percent had disease activity and 57% (n = 25) received immunosuppressive treatment. There was no significant difference in QFN interferon-γ response between subjects with or without disease activity (p > 0.4) and between treated vs non-treated patients (p > 0.5). At follow-up using the Danish tuberculosis registry, there was no occurrence of tuberculosis among study participants."

So it is anti-TB treatment: isoniazid (INH) for nine months. One would think that after hundreds of years of TB we would have a reliable test to determine if someone has TB, latent or otherwise. Nope. Maybe someday.

Rationalization

Slater, M., Parsonnet, J. & Banaei, N. Investigation of false-positive results given by the QuantiFERON-TB Gold In-Tube assay. Journal of clinical microbiology 50, 3105-3107 (2012).

Milman, N., Soborg, B., Svendsen, C.B. & Andersen, A.B. Quantiferon test for tuberculosis screening in sarcoidosis patients. Scandinavian journal of infectious diseases 43, 728-735 (2011).

Gupta, D., Kumar, S., Aggarwal, A.N., Verma, I. & Agarwal, R. Interferon gamma release assay (QuantiFERON-TB Gold In Tube) in patients of sarcoidosis from a population with high prevalence of tuberculosis infection. Sarcoidosis, vasculitis, and diffuse lung diseases : official journal of WASOG 28, 95-101 (2011).

Vermin Bumps

As I have mentioned before, I am an Occam kind of guy. Sing it long and sing it loud: *Numquam ponenda est pluralitas sine necessitate.* But sometimes you have to follow Hickam's dictum when you can't tie everything together into one neat package.

The patient has long-standing rheumatoid arthritis, and is on methotrexate and Rimicade. That's Medicaid for Blackberry users, and boy are they going to need it. Yes, I know it is Remicade, but the obscure pun doesn't work with correct spelling.

He has a pair of kittens, recently adopted and mostly indoor animals, which like to scratch him. This time the little vermin give him a real good mauling of the hand, and he gets cellulitis and lymphangitis. He sees his primary care physician, who gives him a course of Augmentin®. The cellulitis and lymphangitis resolve, but after several days the patient has high fevers, feels awful, and comes to the ER, where his white cell count is zero.

The infection is gone, except at each of the scratches are growing firm, red bumps. He is admitted, started on antibiotics for neutropenic fevers, and they call me. The only other finding of note is neck adenopathy and a mild stomatitis.

So how to put it together?

With the help of the PubMeds, I think the neutropenia is from the Augmentin®. I can' t find any cat-related neutropenias in humans. He had been on a long course of Augmentin® in the past, so was sensitized for what is likely an immune-mediated neutropenia. Not a common reaction by any means, but no other cause was found, and the white cell count returned promptly after stopping the Augmentin®.

The red bumps? Each one is over a cat scratch. Red bumps could be atypical *Mycobacteria*, *Cryptococcus*, *Nocardia* , or *Sporothrix*, but it occurred too rapidly for all of these. I wondered about bacillary angiomatosis (BA) since it looks just like the lesions I saw back in the day with AIDS. Back to the Googles.

Remicade? Nothing. Methotrexate? And bingo was his name-o. There are a smattering of BA cases in allegedly normal people, including one patient on long-term methotrexate. I did not biopsy the bump; I suppose I should have. And cats get stomatitis from *Bartonella*, although this is not mentioned as a symptom in humans.

We, the royal we, will see if he gets better on doxycycline (he did). But I am blaming it all, directly and indirectly, on *Bartonella*.

Rationalization

Desgrandchamps, D. & Schnyder, C. Severe neutropenia in prolonged treatment with orally administered Augmentin (amoxicillin/clavulanic acid). Infection 15, 260-261 (1987).

Kreitzer, T. & Saoud, A. Bacillary angiomatosis following the use of long-term methotrexate therapy: a case report. The West Virginia medical journal 102, 317-318 (2006).

Poll Results

The other white meat is

- pork. 28%
- cat. 25%
- albino anything. 11%
- vampire. 8%
- steak-shaped rice. 19%
- Other Answers 8%
  - -tofu.
  - -HUMAN !!

Wrong. Big Time.

I WAS wrong. Big Time.

Remember the case of fever I declared not due to infection? I thought the fever was due to the myocardial infarction? Boy was I wrong.

His pre-op fever workup was entirely negative, as was the history and physical, for infection, so he was off to the OR for a repair of his aortic valve. What did they find in the OR? A 2 x 3 cm mass in/on the aortic valve. It was the size of a strawberry, the surgeon said. Looks like endocarditis. Where it was on the pre-op echocardiogram is a mystery.

A gram stain of the mass had only one or two white cells and no organisms. I started antibiotics, since if there is an infection I do not want to delay therapy. I wait for the pathology and the patient to wake up enough to give a history and leave for three days. My partner sends the usual culture-negative endocarditis serology. I get back from the long weekend and retake a history: nothing. Zip. No animals, no trauma, no nothing. No reason for culture-negative endocarditis, so I presume a fastidious *Streptoc-coccus*.

Cultures? After five days, one colony of *P. acnes*. I am really uncertain about this beast; it is an extremely rare cause of native valve endocarditis, and the only bicuspid infections on PubMed are teeth, not valves.

Then the pathology comes back: endocarditis, and all stains are negative for organisms.

Then, for the first time in twenty-five years, the serology comes back positive in a culture-negative endocarditis: the *Bartonella henslae* IgM is 256, and the IgG is > 1:2048. Damn. It is probably my first case of *Bartonella* endocarditis. And my partner made it. I am going to get repeat stains/PCR on the vegetation, but that will take a while to get back, and in the meantime, he is on doxycycline. He still denies cat exposure, even though *B. henlsae* is the causative agent of cat scratch disease.

I am wrong all the time. Many of my best diagnoses have been falsified by reality. Looking back over the chart, I still do not think I missed anything. Doesn't help. There is being wrong when you think of a good diagnosis and it does not pan out, and there is being wrong by missing a diagnosis. I am not a fan of the latter.

Rationalization

Fournier, P.E. et al. Comprehensive diagnostic strategy for blood culture-negative endocarditis: a prospective study of 819 new cases. Clinical infectious diseases : an official publication of the Infectious Diseases Society of America 51, 131-140 (2010).

Sohail, M.R., Gray, A.L., Baddour, L.M., Tleyjeh, I.M. & Virk, A. Infective endocarditis due to Propionibacterium species. Clinical microbiology and infection : the official publication of the European Society of Clinical Microbiology and Infectious Diseases 15, 387-394 (2009).

Cilla Eguiluz, G., Montes Ros, M., Lopez Garcia, D., Iraola Sierra, B. & Aramburu Soraluce, V. [Bartonella henselae endocarditis. Report of a case and review of the literature]. Anales de medicina interna 18, 255-258 (2001).

PollResults

I

- don't make mistakes. 0%
- redefine the problem so I am actually correct. 17%
- am frequently in error, never in doubt. 21%
- am plagued by doubt even when I know I am right. 50%
- never commit, so I am never wrong. 8%
- Other Answers 4%

  -am luckier than would be possible in a just universe.

Mindsets

I KEEP two concepts in mind when I see a patient. The first is that everyone in the case has it all wrong. That is almost always not the situation, but I like to form my own ideas as to the diagnosis. The other is to pay attention to the clinical features that do not fit the pattern. Early in medical training you try to tie everything together and tend to ignore outliers that do not belong in the diagnosis. Over the years, I have come to focus more on the outliers.

The patient is my age (old if you are a medical student, young if you are retired) and she is admitted with right-sided abdominal pain and is septic: fevers, hypotension, and so on. Her white blood cell count was 45,000 cells per microliter—that's an impressive leukocytosis. Evaluation for a focus of infection finds an iliopsoas abscess appearing on the CT scan. Usually, iliopsoas abscesses come in two flavors in the Northwest: *S. aureus,* often MRSA, from transient seeding after a "groin pull" (like a taffy pull, but not as much fun); and mixed flora from a colon leak. Tuberculosis and *Coccidiomycosis*, of which I saw more than one in my Los Angeles County fellowship days, are not common here in Portlandia.

They put a drain in it and . . . gram positive branching rods! Actinomycosis!?! That is worth some extra punctuation. She had a sigmoid colectomy four years prior due to strictures and diverticulitis, and the actinomycosis was credited to that intervention, festering on a suture perhaps. And it also grew a touch of *Bacillus fragilis.* For the next several days the patient gets hemodynamically better, but the leukocytosis does not go down. The outlier, as far as I am concerned, is the white cell count. No way is a piddly actinomycosis abscess going to cause this sort of leukocytosis. She must, must I say, have some anatomical issue with her colon besides the diverticulosis seen on CT.

Surgery was hesitant to take her to the operating room without more compelling evidence of a reversible anatomical problem, and the gastrointestinal specialists did not want to colonoscope her for fear of worsening any diverticular process, if there, by inflating her colon with air. "Pop" is not the sound you want to hear in the GI lab. So we waited, and she slowly improved.

Slllllllooooooooowwwwwwwwwwwwlllllllllllyyyyyyyy.

Then she started to bleed, and when GI finally scoped her to find a source, the right colon was ischemic. Is that the cause of the abscess? Or the result of the hypotension? I don't know. I am inclined to credit the colitis for the leukocytosis, but the *Actinomyces* as well? Seem a bit of a stretch. The abscess was 3 x 4 inches in size, so it had been there a while. Disease occurring several years after colonic surgery is not unheard-of, so maybe it was left over from the colectomy.

In the old days *Actinomyces* was more common in men, but in the modern era the classic cause of abdomino-pelvic *Actinomyces* is an IUD, which she never had. I saw one uterine *Actinomyces* years ago as a fellow. It also presents like malignancy and as a complication of appendicitis, but her appendix had been removed decades before. Having concomitant bacteria, especially anaerobes, is the rule with *Actinomyces,* so the patient will end up on a long course of penicillin and metronidazole.

I am not entirely happy with the sequence of events, but as long as the patient improves, that is fine by me.

Rationalization

Choi, M.M., Baek, J.H., Lee, J.N., Park, S. & Lee, W.S. Clinical features of abdominopelvic actinomycosis: report of twenty cases and literature review. Yonsei medical journal 50, 555-559 (2009).

Poll Results

When seeing a patient for the first time

- I assume that all of the prior evaluations are flawed. 7%
- I assume that all of the prior evaluations are fine. 2%
- When you assume, well, everyone knows that old saw. 15%
- I'm so lost in the data dump of the EMR, I have no clue what is going on anymore. 24%
- Some combination of these assumptions. The American Board of Internal Medicine has taught me to pick "all of the above" when available. 49%

....

Bezoar

THE patient is an elderly Cambodian female who comes in with acute abdominal pain and is diagnosed with small bowel obstruction (SBO, one of the less popular cable networks). She doesn't improve with conservative therapy, so off to the OR where the cause of the SBO is a bezoar. It is removed, and the patient improves, and when the pathology returns, they call me.

It is a mass of mold that looks to be *Aspergillus*.

Huh?

A quick PubMed search reveals no cases of fungal bezoars as a cause of SBO, although there are 374 hits in the PubMeds of SBO from a variety of bezoars, mostly hair and vegetable.

The word bezoar comes from the Persian "protection against poison." I am not so sure that is the case.

A history, through an interpreter, reveals that the patient has a particular passion for eating dehydrated mushrooms imported from Asia, and eats nothing else of interest.

There are several case reports of SBO from mushroom bezoars, including one referred to as "Chanterelle intestinal ileus," although I do not know the exact kind of mushroom consumed. Shit take? (Credit to my wife for that one).

But the mold? *Aspergillus* is found in the compost in which mushrooms are grown:

> *"Microorganisms in spent steamed mushroom compost and its dust were enumerated, and identified....The most common bacterial isolate was Bacillus licheniformis. The most common actinomycete isolates were Streptomyces diastaticus and Thermoactinomyces vulgaris. Other actinomycete isolates included Streptomyces albus, Streptomyces griseus, Thermoactinomyces thalpophilis, Thermomonospora chromogena, and Thermomonospora fusca. The most common fungal isolates were Aspergillus fumigatus and Humicola grisea var. thermoidea. Other fungal isolates included Aspergillus flavus, Aspergillus nidulans, Aspergillus terreus, Aspergillus versicolor group. Chrysosporium luteum, Mucor spp., Nigrospora spp., Oidiodendron spp., Paecilomyces spp., Penicillium chrysogenum, Penicillum expansum, Trichoderma viride, and Trichurus spp."*

There is the occasional disease reported in mushroom workers, primarily allergic pneumonitis.

So I hypothesize that it was a mushroom bezoar and that the *Aspergillus* was simply along for the ride.

This is such a fun job. I remain amazed they pay me to do it.

Postscript

After I wrote this, I presented this case to someone who knows more mycology than I do, and was told that I did not have to credit *Aspergillus* as the cause—a mushroom would look just the same on a fungal stain. It was probably mushroom all the way down.

Rationalization

Champeau, M. ["Chanterelle" Intestinal Ileus]. Concours medical 87, 46 (1965).

Arnal-Etienne, S. & Puttemans, T. Small bowel obstruction due to a mushroom bezoar. JBR-BTR : organe de la Societe royale belge de radiologie 92, 114 (2009).

Yoshida, K. et al. Hypersensitivity pneumonitis of a mushroom worker due to Aspergillus glaucus. Archives of environmental health 45, 245-247 (1990).

Kleyn, J.G. & Wetzler, T.F. The microbiology of spent mushroom compost and its dust. Canadian journal of microbiology 27, 748-753 (1981).

Good Times Are Slowly Fading Away

THE patient has a fever, chills, and is completely nonfocal except for a right-sided headache. The erythrocyte sedimentation rate, a marker of inflammation, is very high: greater than 100 millimeters per hour when normal is less then 20. But the rest of the labs are normal, and the primary care physician is walking down the temporal arteritis road. As would I.

However, the patient is in his midfifties, a bit young for temporal arteritis, so as part of the FUO evaluation a CT of the chest/ abdomen/ pelvis is done. The house staff call it a pan-scan. The first CTs came online when I was an intern, and I remember when each pixel was a box about the size of an "o" and there was one CT scanner in the city. Now it is common enough that to scan everything it has its own name. How things change.

What they found was a 4 x 5 inch liver abscess. No right upper quadrant pain, no transaminitis, and no good reason for a liver abscess: no biliary or colonic disease, or bacteremia or anything. The patient's life is duller than mine.

Interventional radiology put a drain in the abscess, and there is pus with gram negative rods. I suspected a *Bacillus fragilis* or some such, but it grew a *Klebsiella pneumoniae*. I should have figured. When I called the lab and asked if it were particularly snotty on the plate, they said yes. It had a positive string test: colonies formed a long string when touched with a loop. You can get an idea what that bacterial snot looks like in the paper in the references by Rivero et al.

Some *Klebsiella* are hypermucoviscous. Bleck. They are like snot on a plate. These hypermucoviscous *Klebsiella* like to form liver abscesses, most often in middle-aged men with no risk factors. I have noticed that for most diseases I have recently slipped into an at-risk demographic, and liver abscess is on the list.

The nonfocal focal infection seems to be the rule:

> *"Patients often complain of vague constitutional symptoms, such as fever and fatigue. Only one-half of patients present with more specific clinical clues as right upper quadrant abdominal pain, jaundice, and hepatomegaly."*

Primarily reported in Taiwan, there are now cases in the United States—a failure, I suppose, of Homeland Security. They need to do more complete cavity searches on people entering the United States, looking for virulent and resistant bacteria.

The extra capsule in the hypermucoviscous organisms increases their virulence by inhibiting killing by white blood cells and complement. I looked into getting it tested for capsular type, but would cost money. In these tight economic times, spending money to satisfy an intellectual itch isn't done. Them were the good old days, and they are fading fast when I was not at risk for disease, and I could order any damn test I wanted because I was curious.

Rationalization

Lederman, E.R. & Crum, N.F. Pyogenic liver abscess with a focus on Klebsiella pneumoniae as a primary pathogen: an emerging disease with unique clinical characteristics. The American journal of gastroenterology 100, 322-331 (2005).

Fang, C.T. et al. Klebsiella pneumoniae genotype K1: an emerging pathogen that causes septic ocular or central nervous system complications from pyogenic liver abscess. Clinical infectious diseases : an official publication of the Infectious Diseases Society of America 45, 284-293 (2007).

Vila, A. et al. Appearance of Klebsiella pneumoniae liver abscess syndrome in Argentina: case report and review of molecular mechanisms of pathogenesis. The open microbiology journal 5, 107-113 (2011).

Rivero, A., Gomez, E., Alland, D., Huang, D.B. & Chiang, T. K2 serotype Klebsiella pneumoniae causing a liver abscess associated with infective endocarditis. Journal of clinical microbiology 48, 639-641 (2010).

A Sigh. Then Diagnosis or Two.

SIGH.

"Parasites" on the clinic schedule. That cannot be good. Parasites never, or rarely, are actual parasites. It is usually some other problem misdiagnosed by the patient or delusions of parasitism.

The patient has a sheath of papers from the referring provider, a naturopath. So I knew that the diagnosis of parasites was likely fanciful. Anyone who suggests homeopathy as a therapy is separated from reality as I understand it. And the labs were the usual uninterpretable nonstandardized serologies and saliva tests.

Double sigh.

If the patient has bought into the fantasy world of alternative diagnosis and treatment, then it may be a contentious interaction. I hate telling patients that all of the money and time they have spent has been, from my perspective, worthless. I hope that I can avoid being a jerk and stick to medicine.

Still, I need to keep an open mind that there is some process going on that has been misdiagnosed as parasites. Once more unto the breach, dear friends, once more.

Turns out that the patient had no exposure history or signs or symptoms to suggest a parasite of any kind. No surprise there. Mostly irritable bowel syndrome symptoms with no reason for IBS such as a prior bacillary diarrhea. The patient had a course of albendazole (maybe—he cannot remember anything but the bendazole part) and a course of a homeopathic antiparasitic nostrum. Neither did a thing. What a surprise. I would expect an imaginary medication to have no effect on a nonexistent disease. I didn't say that to the patient. But really. Using homeopathy is a sure sign that the practitioner doesn't know a burro from a burrow.

A stool screen for ova and parasites was negative from a trusted lab, so I told the patient that I did not think he had to worry about parasites and needed no further diagnostic testing or treatment. Maybe he needed a GI evaluation, not an ID doc, and unless a parasite or *H. pylori* was found on biopsy, he needed no antibiotics of any kind.

But there were two interesting findings. One is that he had telangiectasias (spider veins) and Reynaud's syndrome (fingers turning white in the cold). These two together make up 2/5 of a CREST syndrome, which is a multisystem connective tissue disorder characterized by Calcinosis (thickening of the skin with the appearance of calcific nodules), Reynaud's, Esophageal dysmotility, Sclerodactyly (skin tightening on the fingers), and Telangiectasia. I do not think his GI complaints were esophageal, but he was scheduled to see a rheumatologist for his Raynaud's.

Even more interesting was his phenotype: Abe Lincolnesque. Tall, thin, high-arched palate, big, spidery hands, sternum an inny *(pectus excavatum)*, and an aortic insufficiency murmur. Further questioning revealed a brother who had heart surgery in his fifties; the details were uncertain due to familial estrangement. Sure looks like Marfan's syndrome to me. Marfan's is a genetic connective tissue disorder with an autosomal dominant pattern of inheritance, so family history is usually significant.

I told the patient that once I wander out of ID my knowledge is more general, but that I thought for sure he had Marfan's and, perhaps, CREST syndrome. Odd to find two rare diseases in the same patient, neither diagnosed before, but I am the first MD he as seen in decades. He needed to see a real doctor, that is, not me, to evaluate for both, which he would do and let me know. He didn't. He seemed satisfied he had no parasites and was interested in the other diagnostic suggestions.

I try and keep my eyes open when I see patients. Not everything is infectious (unfortunately) and occasionally I will diagnose something outside my area of expertise.

0 for 2

PART of the fun of the job is calling the bug before the cultures come back. Part of the humility of the job is how often I am wrong. Still, I would prefer to try to predict the infection rather than blather on with a differential diagnosis yet never commit, which ID doctors often do. A year ago I saw a series of acute bacillary diarrheas and I was zero for three in predicting the organism.

First up with this week was a thigh cellulitis in a patient with new-onset diabetes, an admitting blood glucose level of 600 mg/dL when normal is 100 or less. The patient was not that ill, with lots of erythema and a lot of pain, and before the CT scan I said it looked like Group A *Streptococcus* cellulitis. As I went out the door I looked at the CT and the psoas muscle was edematous. Not so sure it was *S. pyogenes*, as I told the house staff, since most pyomyositis is due to MRSA, at least here in Portland. The literature, to my surprise, suggests otherwise. Pyomyositis is more often MSSA:

> *"While MRSA has emerged as the foremost cause of SA infections in a majority of clinical conditions, in this series most patients still had methicillin-sensitive SA as their cause of pyomyositis"*

Necrotizing myositis is increasingly due to MRSA, and in the OR the patient had a necrotizing myositis, so it should have been MRSA. It wasn't. MSSA.

0 for 1.

Next day I was asked to see a patient with diffuse erythema of the leg and a *Streptococcus* of some sort in the blood. Well, I would bet it is Group G *Streptococcus*, since in one series it was three times more common than Group A *Streptococcus* for bacteremic cellulitis, although that was in Finland, and this is not Finland.

> *"Group G Streptococcus (Streptococcus dysgalactiae subsp. equisimilis) was found most often and was isolated from 22% of patient samples of either skin lesions or blood, followed by group A Streptococcus, which was found in 7% of patients"*

Nope. Group B *Streptococcus*. Didn't expect that, especially in a patient with zero risk factors:

> *"Group B streptococcal bacteremia occurred in adults with various underlying diseases, including diabetes mellitus, liver disease, peripheral vascular disease, and hematologic disease, and those*

receiving long-term steroid therapy. Infections causing group B streptococcal bacteremia in adults included decubitus ulcers, pneumonia, endocarditis, cellulitis, arthritis, osteomyelitis, and meningitis."

0 for 2.

Oh well. Babe Ruth led the league in home runs and strikeouts. If you don't swing at the ball, you will never get a hit. Next time I will hit it out of the park, and no one ever remembers the whiffs.

Rationalization

Burdette, S.D. et al. Staphylococcus aureus pyomyositis compared with non-Staphylococcus aureus pyomyositis. The Journal of infection 64, 507-512 (2012).

Changchien, C.H. et al. Retrospective study of necrotizing fasciitis and characterization of its associated methicillin-resistant Staphylococcus aureus in Taiwan. BMC infectious diseases 11, 297 (2011).

Siljander, T. et al. Acute bacterial, nonnecrotizing cellulitis in Finland: microbiological findings. Clinical infectious diseases : an official publication of the Infectious Diseases Society of America 46, 855-861 (2008).

Gallagher, P.G. & Watanakunakorn, C. Group B streptococcal bacteremia in a community teaching hospital. The American journal of medicine 78, 795-800 (1985).

If You Are Not Part of the Solution, You Are Part of the Precipitate

I SEE a fair number of formal consults; they are why my tuxedo is tax deductible. I probably hear of a lot more cases in curbsides, phone calls, and attending noon reports.

The residents presented a case of hemoptysis in a patient with progressive, cavitary pneumonia that, over time, developed the "crescent sign" on the CT scan. The crescent sign is a sure sign of *Aspergillus*, but which one? The patient had a biopsy of the lesion that had a mold that sure looked liked *Aspergillus*, but, as is often the case, it was not sent for culture. Sigh.

If it looks like a duck, and it quacks like a duck, and it flies like a duck, it is probably a duck, but it could be one ugly robin with a bad cough. I like to grow the bugs to clinch the diagnosis.

The pathology, besides having *Aspergillus* looking fungal elements (fungal elements should be called *Aspergillium* I suppose), had calcium oxalate crystals.

You have to go back to 1975, when I was still in high school, to find the first reference of local deposition of calcium oxalate in *Aspergillus* fungal balls.

> *"Presumable oxalic acid produced by the fungus reacts with calcium ions in tissue fluids or blood and is precipitated at calcium oxalate".*

You have to go back to 1891 to find the discovery that oxalic acid is made by *Aspergillus niger* fermentation.

Almost all of the case reports are with *A. niger.*

Oxalic acid may be an inadvertent virulence factor:

> *"Calcium oxalate crystals can induce cellular injury by several mechanisms, and there is increasing evidence that oxalosis-induced tissue damage may occasionally lead to a poor clinical outcome"*

since

> *"In Aspergillus niger, (oxalate) may occur as an unwanted by-product of citric acid fermentation, which, because of its toxicity, must be completely removed"*

A hint with a lack of cultures that the pathogen is *A. niger.*

Who cares? Well, if you are reading this book, you. It is the little things that are often cool. The curiosities in ID go from the very big to the very small.

Rationalization

Kurrein, F., Green, G.H. & Rowles, S.L. Localized deposition of calcium oxalate around a pulmonary Aspergillus niger fungus ball. American journal of clinical pathology 64, 556–563 (1975).

Pabuccuoglu, U Aspects of oxalosis associated with aspergillosis in pathology specimens. Pathology, research and practice 201, 363-368 (2005).

Kubicek, C.P., Schreferl-Kunar, G., Wohrer, W. & Rohr, M. Evidence for a cytoplasmic pathway of oxalate biosynthesis in Aspergillus niger. Applied and environmental microbiology 54, 633-637 (1988).

Yoshida, K. [The correlation between tissue injury and calcium oxalate crystal production in rat's lung with experimental Aspergillus niger infection]. Kansenshogaku zasshi. The Journal of the Japanese Association for Infectious Diseases 72, 621-630 (1998).

Poll Results

I think

- no fact is uninteresting. 52%
- I need know nothing as long there are the Googles. 10%
- facts are uninteresting. I prefer my opinion. 2%
- I prefer to feel that something is true, not to know it is true. 0%
- I would prefer to kick your butt in Trivial Pursuit. 30%
- Other Answers 6%

    -This is a particularly interesting fact!

Old Habits In the New Medical Order

I AM getting too damn old. Rather than responding with a knowing smile at the foibles of medicine, I find I am getting increasingly irritated. It is a sure sign that I am getting to the end of my career.

Maybe it is my increasingly discomfiture with the feeling that I never really know what is going on with patients. People can respond to a perceived lack of control with many kinds of dysfunctional behavior, including irritation. Every other year I have to attend a malpractice prevention seminar as part of carrying malpractice insurance, and a lawyer mentioned in passing that the three significant cases she was defending this year concerned misreading the electronic medical records.

I believe it. We are now all using the Epic software, and in the torrent of tiny print and unneeded detail, I have lost the ability to follow the narrative of the patient's illness in the chart. Given the ability of health care workers to personalize their progress note with volumes of imported garbage in any order they want, it is safe to say that after spending an hour reading the chart, I still feel uncertain that I understand the course of the patient's disease. The EMR is for everything BUT communicating in a meaningful way about the patient.

I think there should be two notes for every patient: the computer-generated note to document all of the data you looked at to justify billing, and a separate note that explains what is going on and why. The latter should take about a paragraph. I spend a lot of time on computers, and there are real benefits to the EMR, but comprehending the patient s disease course is not on the list.

One of the issues with the EMR is that the x-ray report and the film are available, but most of the time I get the sense that no one actually looks at films, just the report.

I get a call that a patient, who had a prior history of being treated for TB, has a chest x-ray that has findings of active tuberculosis, and can I be of some help at sorting it out.

I said put the patient in isolation, and when I get in (I was at another hospital on a different EMR) I would look at the film and see the patient.

Two hours later, I make it to the next hospital and pull up the x-ray. No active TB, just old scarring of the right middle lobe pulling the lung over. I knew what had happened: no one looked at the x-ray, and it was a transcription error.

So I went to radiology, found the radiologist, and while I pulled up the film, asked him where he had learned to read chest x-rays—the school for the blind? It was with a kidding tone of voice; he knew I was bullshitting him.

Sure enough, I was right. He looked at the film and said there was no active TB. I pointed to the dictation. No "no" in the conclusion. A nice string of swear words followed.

Bad transcription, followed by no one looking at the x-ray either independently or with radiology. While no harm was done, the patient did get moved to ICU for three hours into our only available isolation bed.

I have said it before, and I will say it again: always look at the film, not just at the report. And if you do not know why the conclusion on the report describes the findings on film (I know, the pixels, not film) talk to the radiologist.

Several of the radiologists lamented the lack of interaction on complex cases and the lack of clinical history to help guide the interpretation.

That is the future of medicine: too much data to consume, organized in a way that can't be understood, and not bothering to talk with others since it is all on the computer (the EMR is the medical Facebook and Twitter).

See? Old and grouchy.

Postscript

As a further aside, I note I am losing my ability to handwrite. When I take pen in hand, I can no longer write for prolonged periods of time, and it takes an effort to make the pen move as needed; I am losing penmanship muscle memory. The problem is, I type worse than I write.

Poll Results

I cannot grok

- the EMR. 14%
- Twitter. 21%
- Facebook. 5%
- the word "grok." Say what? 46%
- anything after 1999. 11%
- Other Answers 2%

    -If they don't grok, then fuggetaboutem.

Mystery

I HAVE been doing ID for twenty-five years now if you include fellowship, and as I have alluded to in the past, it is still the most interesting part of medicine. Ever.

But after all this time I still cannot understand the process by which I generate a diagnosis. It is a mystery how I have internalized information, and as I take a history it is aggregated, analyzed, and ideas as to the cause of a patient's illness pop from somewhere into consciousness.

I was on call this weekend and was asked to see a patient who the hospitalist could not figure out. This is never a promising consult, as I work with an outstanding bunch of hospitalists who can usually beat me to the diagnosis.

A middle-aged male gets a boil on the cheek. He is seen in the ER, it is lanced with more blood than pus expressed from the lesion, and he is sent home on cephalexin. From there, things get worse.

Over the next twenty-four hours, he develops high fevers, conjunctivitis, submandibular lymphadenopathy and red, tender bumps mostly on the arms. When I see them, they look primarily like *erythema nodosum* (medical speak for "red bumps"), although I am not so great at making rash decisions.

I run through my usual history and exposures, and the only thing he mentions is cats. A bell goes off in my head. "Bing." Or for me, "Google." Unbidden, the phrase *Parinaud's syndrome* pops into my brain. Somewhere, somehow I remember that cat scratch causes conjunctivitis, and it is Parinaud's syndrome. Or am I thinking of Pernod? Both are noxious. I am not entirely certain and this makes me glad I practice in the Internet era. The first hit is Parinaud's syndrome, also known as dorsal midbrain syndrome. Nope.

But there is Parinaud without the "s," an oculogladular form of cat scratch disease. Yep. Fevers, lymphadenopathy, and conjunctivitis. And there are rare cases of *erythema nodosum* reported with cat scratch disease because the lesions do not look like the typical red bump disease, bacillary angiomatosis, seen with cat scratch disease.

The cause of cat scratch disease is *Bartonella,* diagnosed by PCR. The test is pending, and I will have a negative result in a week or three. I bet it is all a reaction from the cephalexin, although I can't find reports of a syndrome like this with cephalexin. Adverse drug reactions are so much more common than cat scratch disease. So we will see.

Rationalization

Jawad, A.S. & Amen, A.A. Cat-scratch disease presenting as the oculoglandular syndrome of Parinaud: a report of two cases. Postgraduate medical journal 66, 467-468 (1990).

Sarret, C. et al. [Erythema nodosum and adenopathy in a 15-year-old boy: uncommon signs of cat scratch disease]. Archives de pediatrie : organe officiel de la Societe francaise de pediatrie 12, 295-297 (2005).

Poll Results

The best exposure for odd infections remains

- cats. 16%
- dogs. 0%
- children. 25%
- food. 2%
- travel. 55%
- Other Answers 2%
    - -parasites.

Life List

INFECTIOUS diseases is like birding, except interesting. I do enjoy what birds have to offer, although I prefer the taste of condor to spotted owl. I wonder what dodo would have tasted like. But I digress. Birders keep a list of birds they have seen, a life list, as do I with infectious diseases—although my list shows up here, not in a spiral notebook.

The number of infections is long, and the great Pacific Northwest is not a hotbed of exotic diseases, but occasionally one comes my way, and I can check it off the list.

The patient is a middle-aged African male who presents with abdominal pain, fevers, and ascites. The CT scan shows massive adenopathy everywhere. The labs showed hypercalcemia, and a lymph node biopsy had a T cell lymphoma. I was asked if the patient had TB or perhaps another granulomatous process that led to the hypercalcemia.

Nope. It is interesting that Vitamin D deficiency and polymorphisms in the Vitamin D receptor are associated with increased risk of TB, and that granulomas make Vitamin D, resulting in hypercalcemia.

> *"Hypercalcaemia has been described in adult patients with most granulomatous disorders, including TB, although the majority are asymptomatic. Rates of hypercalcaemia in TB in adults vary widely from 6 to 48%."*

But that is not the cause of the hyperkalemia. There were no granulomas on the biopsy. I think it is the lymphoma, a specific lymphoma due to HTLV -1, and today the serology came back positive for HTLV-1.

> *"Some 10-20 million people world-wide are infected with HTLV-1, which is endemic to Japan, the Caribbean, Central and South America and Africa where the seroprevalence is 1-20%. The estimated lifetime risk of developing ATLL in a person positive for HTLV-1 is 2-5%, with a latency of 20-30 years. . . .Hypercalcaemia occurs in about 50% of patients with HTLV-1 induced*

ATLL. . . . The mechanism of hypercalcaemia associated with ATLL seems to be stimulation of bone resorption by the production of parathyroid-hormone-related peptide (PTH-rP), with ATLL cells constitutively expressing a large amount of PTH-rP mRNA."

Weird. HTLV-1 lymphomas makes lots of parathyroid hormone. I wonder why. Does it help the virus somehow? It must. The other disease caused by HTLV-1 is tropical spastic paraparesis and myelopathy, which is even more dreadful than lymphoma.

That is the one enormous advantage birders over ID doctors have: no human suffering, outside of their own with neck spasms from looking skyward, is required to complete their life list.

Rationalization

Davies, P.D. & Grange, J.M. Factors affecting susceptibility and resistance to tuberculosis. Thorax 56 Suppl 2, ii23-29 (2001).

Edwards, C.M., Edwards, S.J., Bhumbra, R.P. & Chowdhury, T.A. Severe refractory hypercalcaemia in HTLV-1 infection. Journal of the Royal Society of Medicine 96, 126-127 (2003).

Payne, H.A., Menson, E., Sharland, M. & Bryant, P.A. Symptomatic hypercalcaemia in paediatric tuberculosis. European respiratory review : an official journal of the European Respiratory Society 20, 53-56 (2011).

Gessain, A. & Mahieux, R. Tropical spastic paraparesis and HTLV-1 associated myelopathy: clinical, epidemiological, virological and therapeutic aspects. Revue neurologique 168, 257-269 (2012).

Poll Results

I keep a life list of

- infections. 18%
- birds. 0%
- beers. 9%
- all those who have wronged me and they WILL pay; oh how they will pay. 55%
- people I have crushed and humiliated in my rise to the top. 9%
- Other Answers 9%

    -excellent sushi joints.

Say It; Don't Spray It

So often in ID it is "close but no cigar."

Everything points to a diagnosis, but the damn bug doesn't grow. It's there on the gram stain, but prior antibiotics or the host immune system finishes it off, and I am close, oh so close to a definite answer. It is the smell of the bread without getting a taste.

The patient has bacterial meningitis. Fevers, headache, increased white blood cells in the CSF, high protein, and low glucose levels in the blood. Gram-positive cocci in chains are seen, but nothing grows.

I know what it should be, since this meningitis occurred shortly after a surgical procedure—namely a "blood patch," where the patient's blood is used to patch a hole in the dura mater of the spinal cord. It should be *Streptococcus salivarius*. Most post-procedure (lumbar puncture, myelogram, and epidural anesthetic) meningitis is due to *S. salivarius*, but the damn thing will not grow.

It is an interesting question whose *S. salivarius* it is. When I was a fellow I was taught it was the patient's. I had a case where the woman chewed gum through the entire procedure, and the hypothesis is that the unfortunate confluence of transient bacteremia from gum chewing was combined with a needle in the back resulted in the injection of the bacteria directly into the CSF.

Seems unlikely, but my practice is based on bad luck. I had a patient recently, a retired lawyer, who opined that he did not believe in coincidences. I am just the opposite; I see almost all events as random badness. The book, *The Drunkard's Walk: How Randomness Rules Our Lives,* by Leonard Mlodikow is an excellent review of the topic.

I found a paper that suggested that much of the *S. salivarius* is coming from the medical staff when they do not wear masks during procedures. A fine spray of spittle-laden bacteria wafts into the air and is pushed into the CSF with the needle, often with fatal results. I remember how close my face was to the back when I did lumbar punctures back in the day. I do not know if masks were worn during this procedure.

It is a rare complication, and you are likely to go a lifetime without causing one, but wear a mask anyway when you stick a needle into someone's back.

The patient? Doing fine clinically, although there is the suspicion on MRI of discitis. I have seen that as well.

Rationalization

Shewmaker, P.L. et al. Streptococcus salivarius meningitis case strain traced to oral flora of anesthesiologist. Journal of clinical microbiology 48, 2589-2591 (2010).

Yazici, N., Yalcin, B., Cila, A., Alnay, A. & Buyukpamukcu, M. Discitis following lumbar puncture in non-Hodgkin lymphoma. Pediatric hematology and oncology 22, 689-694 (2005).

Poll Results

I wear a mask

- for all procedures. 19%
- only at Mardi Gras. 19%
- to hide my hideous features. 19%
- to block my foul breath. 3%
- never. Germs are a myth; have you ever seen one? 30%
- Other Answers 11%
    - -only when committing crimes.
    - -major OT procedures.

I Have Returned

SPRING break is over. Bummer. I spend it on the Oregon coast, no television. A couple of years ago I went off for spring break and I returned to find out that H1N1 had erupted in Mexico. As a result, every time I return from time off I fret that some new plague has cut loose in my absence. Not this time. Bummer.

The first day back is always gruesome. I am amazed at how much paperwork can pile up during a one-week absence. And trying to get caught up made me hate the electronic medical record software, Epic, even more. EMR may be the future, but so is global warming and the sun turning into a red giant.

At least the diseases remain interesting, if awful. When I left for vacation, I had a patient who presented with a headache and with decreased mental status. The MRI was read as a probable glioblastoma, but the patient declined rapidly and the diagnosis was found to be a brain abscess.

I have seen a half dozen brain tumors over the years that were not actually tumors, but were atypical presentations for brain abscesses. Atypical in that they had no fever (actually not that uncommon) and no good reason for a brain abscess: neither a bacteremic illness nor an odontogenic source, as was the issue with this case.

In the OR it was pus and gram positive cocci, and I was off for vacation.

I expected a viridans *Streptococcus*, which is what is usually grown in brain abscesses, although there are far more organisms in the pus than can be grown. Using molecular techniques one study

> "... *Identified significantly fewer types of bacteria (22 strains) than did molecular testing (72 strains; P = .017, by analysis of variance test). We found that a patient could exhibit as many as 16 different bacterial species in a single abscess. The obtained cultures identified 14 different species already known to cause cerebral abscess. Single sequencing performed poorly, whereas multiple sequencing identified 49 species, of which 27 had not been previously reported in brain abscess investigations and 15 were completely unknown. Interestingly, we observed 2 patients who harbored Mycoplasma hominis*"

It grew a *S. mitis*, an oral *Streptococcus*. And *Capnocytophaga* species. That I did not expect. Six cases in the PubMeds, some from dog and cat bites—not a surprise, since it is a beast found in the mouths of dogs and cats. And people. We have our own *Capnocytophaga*. It is one of the many reasons why I do not let dogs and cats lick me, besides the fact that they also lick their rear. Ick.

Interestingly, there was one *Capnocytophaga* in the paper, and a case with a viridans *Streptococcus* in a patient with cyanotic heart disease. There was no animal exposure in my case, so I suspect the source is the patient.

Rationalization

Al Masalma, M. et al. The expansion of the microbiological spectrum of brain abscesses with use of multiple 16S ribosomal DNA sequencing. Clinical infectious diseases : an official publication of the Infectious Diseases Society of America 48, 1169-1178 (2009).

Engelhardt, K. et al. Brain abscess due to Capnocytophaga species, Actinomyces species, and Streptococcus intermedius in a patient with cyanotic congenital heart disease. European journal of clinical microbiology & infectious diseases : official publication of the European Society of Clinical Microbiology 21, 236-237 (2002).

Jo, E. et al. Capnocytophaga endodontalis sp. nov., Isolated from a Human Refractory Periapical Abscess. Current microbiology 75, 420-425 (2018).

Poll Results

Welcome back. While you were gone

- I didn't notice. 18%
- my life stopped. 48%
- I re-read *Lord of the Rings*. 5%
- I spent my time on the Widows Walk. 8%
- I read someone else now. 18%
- Other Answers 5%
    - -I had a panic attack.
    - - I had a vacation.

They Say It Is So, But I Am Skeptical

An elderly male with a fever. He cannot give me a history because of a major stroke, and all I know is that he has been in a recent decline and had a day or two of fevers, and is brought in for evaluation.

Past medical history, besides the major stroke, is significant for panhypopituatarism from thirty years ago. He has the usual fever workup and nothing else .

I see him and also find nothing, and declare that the patient, who is otherwise normal in his labs and studies, does not have a bacterial process and to watch and wait and the fevers will either go away, or the reason will become evident. Duh. Like there is any other option.

But sure enough, his fevers were gone the next day. Am I good or what?

It turns out or what.

For some unknown reasons, at an unknown time at the skilled nursing facility where he lives, his thyroid medication had been stopped and his prednisone had been decreased to 1 mg a day. Things happen at nursing homes that are not explainable by science. At the same time I declared that his fevers were noninfectious and would fade, his thyroid medication was restarted, and his prednisone was increased to normal physiologic levels.

The Chief of Medicine saw the patient the next day and suggested that the cause of fevers was adrenal insufficiency. There are reports:

> *Except for asymptomatic bacteriuria in one patient (who remained febrile despite appropriate antibiotic therapy), infection was ruled out in all cases, and fever was attributed to "masked" thyrotoxicosis, triiodothyronine (T3) toxicosis, subacute thyroiditis, primary adrenal insufficiency, secondary adrenal insufficiency and pheochromocytoma.*

I made that diagnosis years ago in a patient, and subsequently forgot about it as a cause of fevers for this case.

Adrenal insufficiency is reported as a cause of fevers, but as I think about it, I am skeptical. The cause of adrenal insufficiency in the old days was often infectious (TB and cytomegalovirus), so was the fever due to the infection that caused the adrenal insufficiency? Because I would think that adrenal insufficiency would cause hypothermia, not a fever.

I cannot find a mechanism for the hyperpyrexia from adrenal insufficiency, and without a reasonable mechanism, it seems a stretch. The only reason I can think of is that poor peripheral perfusion from low blood pressure prevents radiation of heat—so that it would be more like heatstroke as a mechanism.

The few cases on the PubMeds have other reasons for the fever: tumor and infection. But all of the textbooks say something like orthostatic hypotension, fever, and hypoglycemia characterize acute adrenal crisis.

The only real discussion I can find is the association of adrenal hemorrhage with fever is from 1926. Cool article. But nothing of note since.

His blood pressure and electrolytes were curiously fine, so I am skeptical for that reason as well that he was in adrenal crisis.

Oh well. He is better, and I will take full credit.

Rationalization

Simon, H.B. & Daniels, G.H. Hormonal hyperthermia: endocrinologic causes of fever. The American journal of medicine 66, 257-263 (1979).

Cramer, W. Fever, Infections and the Thyroid-Adrenal Apparatus. British journal of experimental pathology 7, 95–110.117 (1926).

Poll Results

Despite a lack of data I believe in

- fevers from adrenal crisis. 4%
- Centers for Medicare and Medicaid guidelines. haha. Sigh. I love a good laugh. 8%
- the results of the cultures. 4%
- peanut butter. But not Peter Pan. Jif is the way to go. 33%
- paranormal phenomena in long-term care facilities. Weird, unexplainable, stuff just happens there. 50%

Stay Away from My Mother

A TRUE story. When I first read it, I thought it was made up to make a point. I did not write the text, but it occurred at a Portland hospital in mid-March and I reproduce it with permission.

Patient Story: Spreading the Flu 4-2012
It starts with Patient #1, a 47 year-old woman admitted through the emergency department (ED) in mid-March with fever and

shortness of breath. She was transferred to an inpatient unit with a mask on, which triggered the staff on the receiving unit to implement droplet precautions. Initially thought to have pneumonia, testing confirmed her symptoms were the result of influenza type A, H1N1. After four nights in the hospital, she was discharged home after an uneventful hospital stay and a flu shot.

Patient #2, next door to Patient #1, is a 61 year-old man who was admitted in early March for a GI bleed with multiple co-morbidities. His progress was steady until nine days after admission, when he developed a new fever and respiratory symptoms. These symptoms developed on the same day of Patient #1's admission. Influenza was suspected two days following the development of his fever, and staff implemented droplet precautions. Lab testing confirmed influenza type A. He remained hospitalized for two more days and received a flu shot before being transferred to a skilled nursing facility.

Down the hall, Patient #3, a 71 year-old man, was admitted two days after Patient #1 for acute stroke and urinary tract infection. On day 3 of his hospitalization, he developed a fever and cough. Lab testing confirmed influenza type A. Droplet precautions were ordered with the lab test for influenza. He remained hospitalized an additional four nights and received a flu shot before being discharged.

Patient #4, a 73 year- old man, down the hall from the first two patients and around the corner from Patient #3, was admitted on the same day as Patient #1 following a fainting event at home. Due to his long-standing heart issues, he was kept overnight for observation and discharged the following morning. However, he returned to the ED three days later with continued symptoms. He was discharged from the ED only to return the next day with shortness of breath. Six hours after being readmitted, staff suspected influenza and ordered droplet precautions. His lab tests returned positive for influenza type A. After spending three nights in the hospital, he was discharged home after receiving a flu shot. The following day, he was admitted to the intensive care unit and continued receiving treatment as an inpatient for secondary pneumonia, a complication of his influenza type A infection.

The fifth person in our story is a nurse on the unit where these four patients were admitted. She works on a nursing unit whose hand hygiene performance is currently 67%, and where 85% of the unit staff were vaccinated for this year's seasonal flu. The particular nurse in this case, however, was 1 of only 9 on the unit who chose not to be vaccinated. Her manager stated that the reason the nurse gave for not receiving the vaccine was that she "was not convinced of the evidence that the vaccine protects patients from transmission … she said she would get the vaccine if she truly believed it protected her patients, but that she didn't."

This nurse cared for Patient #1 on her first day of admission. She cared for Patient #2 on the eighth and ninth day (when he developed flu symptoms) of his stay. She also cared for Patient #3 on the first two days of his inpatient stay. There does not appear to be any direct contact with this nurse and Patient #4. The nurse in our story developed symptoms consistent with influenza three days after working with Patient #1 and Patient #2 (which is the usual 1- to 4-day incubation period for influenza). Due to symptoms, she only worked a partial shift that day. Suspecting her symptoms may be influenza, she used a mask until relief staff was available. She returned home and was able to care for herself without medical intervention. She was not tested for influenza and remained off work for one week.

She is still undecided about receiving the flu vaccine.

It is messy, of course, as the real world often is. And the vaccines are not 100% effective in preventing flu and its spread; neither are seatbelts in preventing death and injury in car accidents.

Belief is what you do when there is an absence of facts.

One of the aggravating issues at the hospital is getting health care providers to get the flu vaccine. The best I can do is convincing about 70% of employees to get vaccinated. Across the country, those numbers are the same, although MD and RN vaccination rates hover around 80%. That is pathetic, since I have the old-fashioned idea that it is incumbent upon us to protect our patients, especially given that nosocomially acquired flu has a

27% mortality rate and people shed the flu before they are ill or likely to come to work when ill:

"...we estimated that 1%–8% of infectiousness occurs prior to illness onset. Only 14% of infections with detectable shedding at RT-PCR were asymptomatic, and viral shedding was low in these cases."

Silent shedders may be unimportant in the community, but if it were a nurse or respiratory therapist going from room to room, it could result in rapid spread of disease.

The anti-vaccine crowd sees low vaccination rates in health care workers as an indictment of the vaccine, suggesting that health care workers know something that the rest of us do not. That is not the case. I have wandered the hospital giving flu shots, and there are the same lame reasons given by the health care workers like the general population, although health care workers should know better since they ostensibly have access to all of the information and the best minds (mine) in the field. I publish an article every October on why health care workers are dumbasses for not getting the flu vaccine. Oddly, people take exception at being called a dumbass.

What factors are important in health care workers getting the flu vaccine?

Knowing that the vaccine is effective (mhRR 2.22; 95% CI 1.93 to 2.54), being willing to prevent influenza transmission (mhRR 2.31; 95% CI 1.97 to 2.70), believing that influenza is highly contagious (RR 2.25; 95% CI 1.66 to 3.05), believing that influenza prevention is important (mhRR 3.63; 95% CI 2.87 to 4.59) and having a family that is usually vaccinated (RR 2.32; 95% CI 1.64 to 3.28) were statistically significantly associated with a twofold higher vaccine uptake.

And infected health care workers markedly increase your risk of influenza-like illness if you are exposed in the hospital.

For patients exposed to at least 1 contagious HCW compared with those with no documented exposure in the hospital, the RR of

HA-ILI was 5.48 (95% confidence interval [CI], 2.09-14.37); for patients exposed to at least 1 contagious patient, the RR was 17.96 (95% CI, 10.07-32.03); and for patients exposed to at least 1 contagious patient and 1 contagious HCW, the RR was 34.75.

Part of the problem is there have been no studies to demonstrate that specifically vaccinating hospital workers will decrease influenza in patients. There are certainly buckets of biologic plausibility to suggest that if I do not get flu because of the vaccine, I will not pass it on to you as I make rounds in the ICU. There have been studies to suggest benefits in nursing homes, but of course nursing homes are not the same as acute care hospitals, where the length of stay is increasingly measured in hours rather than in days, lessening the risk of exposure. The literature now has an extremely suboptimal study that suggests vaccinating health care workers decreases influenza in the hospital.

In summary, our observational study indicates a protective influence of vaccination of more than 35% of HCW on HAI in patients. Other experimentally-designed investigations are needed to demonstrate the effectiveness of HCW vaccination in the control of influenza outbreaks in healthcare settings, and to determine the threshold for vaccinated HCW proportion with more accuracy. Our findings must not be misinterpreted. To date, the HCW vaccination rate of 35% is not optimal to control HAI.

But I will take what I can get, since the more who are vaccinated, the fewer who will get the disease. What is curious to my mind is why people would think otherwise about the vaccine. There is no definitive study that demonstrates that vaccination of health care workers will prevent spread to patients, but the preponderance of data would strongly suggest it should, and given our moral obligation to protect our patients, I would think it would result in vaccination rates of almost 100%.

Since the flu vaccine is not close to 100% effective, there are other measures to help decrease the spread. A career in infection control has consistently reinforced the idea that there is never one intervention that completely decreases the risk of infection, and no one—except the anti-vaccine wackaloons—think that the vaccine is an all-or-nothing phenomenon. Simple mechanical issues often do not demonstrate benefit: proper handwashing and masks are of variable help in outbreak situations if started early.

During the 2009/10 and 2010/11 pandemic seasons, Dr. Buchholz and colleagues compared 84 households in which 1 member had influenza. The households were randomized into 3 groups: facemask use and intensified hand hygiene; facemask use only; and no intervention (control group). Among the 218 noninfected contacts in the 84 households, 35 (16%) developed flu. An intention-to-treat analysis found no benefit from facemasks and handwashing, nor from facemasks alone. However, when the intervention was started within 36 hours of symptom onset in the index case, pooled data from the 2 intervention groups showed an 84% reduction in secondary flu infection. In addition, when facemask alone was considered in a per protocol analysis, there was a 70% reduction in risk.

Other evaluations of the effects of masks and handwashing suggested they were not very good for flu prevention:

Data from 788 households were collected from April 2008 to February 2011. The median age of the household member with flu was 5 years, and 724 (92%) slept in the same room with their parents. Among the household contacts, 31% became infected. Handwashing was found to have no significant benefit, and facemask use was associated with a modest increase in risk for influenza infection.

There is a difference between using a mask and handwashing in the community to prevent acquisition, which is only of perhaps modest benefit, and is probably not the same as masking an infected person to prevent spread in the hospital.

It is doubly interesting since there was a proof-of-concept study that demonstrated that flu infection may be acquired through the eyes. Volunteers were squirted with live flu vaccine while wearing various protective measures, and as long as there was no eye protection, they became infected with the vaccine strain.

My advice to you and yours during the flu season: If admitted to the hospital, you are probably part of a population at risk for dying of influenza, and if your provider has not had the vaccine, they may not be convinced of the contagiousness of flu nor its seriousness. They surely don't give a rat's ass about your safety. Forbid them from entering your hospital room and nicely request another health care worker. Do not let the unvaccinated person into your room. Ever. Put a sign on your door: **No Entry to Unvaccinated People.**

Given that hand hygiene is often not 100%, I would be further disinclined to allow the unvaccinated health care worker in a mask to touch me and mine, never knowing for sure where those hands may have been. Actually, I do know. And you should be ashamed.

I would prohibit the unvaccinated health care worker from entering my room even if masked, although there are some hospitals in my area that are having workers who are not vaccinated wear masks. I should probably wear a mask for aesthetic reasons. I suppose if my unvaccinated provider wore a spacesuit, I would let them in the room.

If anyone from my family is admitted to the hospital, especially my mother, I will not allow anyone in her room who has not had the flu vaccine. I suggest you do the same.

But otherwise, stay away from my mother.

I Forgot

"There is nothing wrong with your browser. Do not attempt to adjust the picture. We are controlling internet. If we wish to make it louder, we will bring up the volume. If we wish to make it softer, we will tune it to a whisper. We will control the horizontal. We will control the vertical. We can rickroll the image, make it flutter. We can change the focus to a soft blur or sharpen it to crystal clarity. For the next 10 minutes, sit quietly and we will control all that you see and hear. We repeat: there is nothing wrong with your browser. You are about to participate in a great adventure. You are

about to experience the awe and mystery which reaches from the inner mind to — Infectious Diseases."

The advantage of writing a book is that I get to control the message. I present the cases that have an answer, or if no answer is available, at least a reasonable explanation as to why there is no answer. Like Oz (the doctor or the wizard) it is easy to give the illusion that I know all and see all. Hardly the case.

The patient is a middle-aged diabetic who had a renal transplant three years ago with zero complications; she is stable on sirolimus, an immunosuppressant that prevents transplant rejection. A month ago she started with a cough. The cough did not remit, and over the next month she had a progressive cough and shortness of breath. That is it. No fevers or other symptoms. She had a Z-pack (oral azithromycin antibiotic) (who doesn't?) with no improvement, and is admitted with hypoxia.

A workup by the hospitalist gives me a normal complete blood count and comprehensive metabolic panel, and a chest x-ray showing bilateral butterfly pattern interstitial infiltrates. She is treated for atypical pneumonia with no resolution, and they call me. Tests for pertussis and flu are negative.

I find nothing further in the history, and the exam shows poor airway movement and crackles. Nothing else.

Well, says the all-knowing ID consultant, it is either PJP (although it would be nice if there were fevers) or sirolimus lung disease, which I prefer given the lack of infection symptoms.

I send off a lactase dehydrogenase test to look for tissue damage, and it comes back at the upper limit of normal (280 U/mL). I also get a 1-3 beta glucan test to indicate fungi, hoping to make the diagnosis of PJP without a bronchoscopy. What are common things? Common. Am I looking for common things? Nope. Nothing even thinking about the usual suspects.

So pulmonary sees the patient and orders a B-type natriuretic peptide (BNP) test. This is a relatively new test that can indicate congestive heart failure. BNP is made in the ventricles, and when the ventricular walls are stretched from heart failure, BNP levels rise. Values below 100 pg/mL rule out heart failure, and a diagnosis is supported at levels above 900 pg/mL. Hers is 3,000.

It is all heart failure. Crap. I wish I had a good reason for not considering the basics. Blame the electronic medical records for making the narrative impossible to follow, or my assumption that the simple things were ruled out before they called me, or the lack of other findings or prior history. It would be nice to blame others, but I got nothing. No harm at least, because I forgot.

> *"Two simple words. Two simple words in the English language: "I forgot!" How many times do we let ourselves get into terrible situations because we don't say "I forgot"? Let's say you're on trial for armed robbery. You say to the judge, "I forgot armed robbery was illegal." Let's suppose he says back to you, "You have committed a foul crime. You have stolen hundreds and thousands of dollars from people at random, and you say, 'I forgot'?" Two simple words: Excuuuuuse me!!"*

Rationalization

Lee, H.S. et al. Sirolimus-induced pneumonitis after renal transplantation: a single-center experience. Transplantation proceedings 44, 161-163 (2012).

Kempainen, R.R., Migeon, M.B. & Wolf, F.M. Understanding our mistakes: a primer on errors in clinical reasoning. Medical teacher 25, 177-181 (2003).

Trying to Change

THE best thing about writing blogs, doing podcasts, and working at a teaching hospital is that all of this helps keep me current in the literature and in medical practice. It is hard to keep up, given the volume of publications (10,000 titles a year in ID alone), and even harder to incorporate the literature into practice.

One of the oddest comments against science-based medicine I have ever read was a commentator who said,

> *"This cobble together thinking that Dawkins spouts is usually what happens when people keep changing their stories to fit the evidence and usually implies that one does not have a grasp of the subject."*

Weird. Changing my story to fit the evidence defines how I try to practice, and is what demonstrates an understanding of a topic. As I get older and more set in my ways, I do find increasing intellectual inertia in changing my practice. It takes more effort to think differently, despite being a Mac user from the beginning.

Staphylococcus aureus pays my mortgage, and it is a beast I think I understand. *S. aureus* bacteremia is tricky since *S. aureus* loves to go elsewhere and set up metastatic complications. Endocarditis and abscesses are what I expect with *S. aureus*.

The patient was admitted to Outside Hospital with a B&B: boil and bacteremia in a diabetic. It was MSSA, she had the usual workup (negative transthoracic echocardiogram and repeat blood cultures) and no complication was found, so she was treated with fourteen days of IV cefazolin. No issue there. Two weeks after completing the antibiotics, she presents with fevers and abdominal pain and workup shows a perinephric abscess/iliopsoas abscess that grows the same MSSA.

I assume that it had been festering all that time and reared its ugly head after the antibiotics were stopped. As a rule, I do not go looking for metastatic infections with *S. aureus* (except the valve) unless the patient has focal symptoms. Occasionally *S. aureus* will cause a "cold" infection, but usually it loves to cause symptoms when there is a pocket of pus that needs draining. Or does it?

Apropos of this case, this month in *Medicine* there was a curious article where the authors went looking for metastatic infections in patients with gram-positive bacteremia. They used trans-esophageal echocardiography and positron emission tomography scanning (which is at least five grand a pop in the United States). These methods found infection in most of the patients, even though most didn't have focal symptoms:

> *"An intensive search for metastatic infectious foci was performed including F-fluorodeoxyglucose-positron emission tomography in combination with low-dose computed tomography scanning for optimizing anatomical correlation (FDG-PET/CT) and echocardiography in the first 2 weeks of admission. Metastatic infectious foci were detected in 84 of 115 (73%) patients. Endocarditis*

(22 cases), endovascular infections (19 cases), pulmonary abscesses (16 cases), and spondylodiscitis (11 cases) were diagnosed most frequently. The incidence of metastatic infection was similar in patients with Streptococcus species and patients with S. aureus bacteremia. Signs and symptoms guiding the attending physician in the diagnostic workup were present in only a minority of cases (41%)."

I find the lack of signs and symptoms and the lack of their utility sobering:

"the number of foci revealed by symptom-guided CT, ultrasound, and magnetic resonance imaging remained low."

Still, I am disinclined to get a PET scan on every patient with gram-positive bacteremia, not that it would be approved or paid for anyway. Although the most dangerous words in medicine remain, "In my experience," the study doesn't ring true—he says, again demonstrating a logical fallacy. But I don't see that many relapsed undertreated gram-positive abscesses, perhaps because I tend to treat the hell out of gram-positive bacteremias. I would have gone four weeks with this patient because of diabetes. But with the next gram-positive bacteremia, am I going to change my story? I don't know. I'll think about it.

Rationalization

Vos, F.J. et al. Metastatic infectious disease and clinical outcome in Staphylococcus aureus and Streptococcus species bacteremia. Medicine 91, 86-94 (2012).

Poll Results

Changing my mind

- is easy. It is why my mind is so clean. 23%
- is more difficult with the passage of time. 17%
- I rationalize the information to give the illusion that I have not changed my mind. 47%
- my opinions are fact-free and require no changing. 3%
- facts are a waste of time to learn; I prefer uninformed ignorance. 7%

- Other Answers 3%

  -To quote the *Star Trek* episodes, "Spock's Brain," "Brain and Brain, and "What Is Brain?" I'm having trouble grasping a concept of mind as a static point. Folly.

The Arrogance of Ignorance

I AM in a ranty pissy mood, so I might as well vent my spleen with a pissy rant. Since it is my book, in the immortal words of Cartman from *South Park,* I do what I want.

"I think," he says, dripping with confirmation bias, "that there is no part of medicine where those who have no clue as to what they are doing are more willing flail about in ignorance than infectious diseases." Boy, that is an awkward sentence.

No one in a major medical center would treat cancer without an oncologist, treat an acute myocardial infarction without a cardiologist, remove an appendix without a surgeon, or deliver a baby without an obstetrician. But people seem more than willing to treat infectious diseases when they have no idea as to what they are doing and without understanding the significance of the infecting organism. And the infection voted most likely to be screwed up is *S. aureus* bacteremia.

The willingness to flail about in ignorance varies.

University hospitals are greater than community hospitals.

Nonmedicine specialties are more than medical specialties.

Residents (sometimes) are more than attendings.

If you are on a university orthopedics trauma service and you get a serious infection, well, may the Lord have mercy on your soul.

My patient was admitted with multiple traumas to an ortho-trauma service at a major university hospital somewhere in Oregon/ Washington/ Idaho/ northern California after a motorcycle accident. Lots of fractures and trauma that required prolonged hospitalization. During the admit she had cellulitis with a MSSA bacteremia, with positive blood cultures three days apart.

Sustained *S. aureus* bacteremia: bad. She was given a brief course of IV antibiotics and then a short course of oral ones. Her workup and treatment for the *S. aureus* bacteremia was, in a few words, totally inept. At no time was someone who knows what they are doing consulted. They preferred to flail about in ignorance.

After being transferred to Portland, the patient had two months of progressive back pain that was not evaluated in ortho-trauma follow-up. Eventually, she ended up in one of my hospitals with an iliopsoas abscess/discitis/spinal osteomyelitis from MSSA. Duh. Wonder where that *S. aureus* came from, he says in his best Lennie voice.

It is not like it is hard to look up the significance of a disease or lab finding in the era of the interwebs. These case histories are in part about my daily exploration of the curiosities of my practice. I see something, I ask "Why?" I look it up and we all learn together. You have to have the intellectual curiosity of a stone not to look up the evaluation and treatment of sustained *S. aureus* bacteremia. Between Up-to-Date and PubMed it should take but a few minutes.

And there are at least three self-serving clinical trials to show that outcomes are improved for *S. aureus* bacteremia with infectious disease consultation, which they didn't bother to do. Oh no, don't actually take the time to call someone with expertise.

Of course, those who most need to read this book are the least likely to do it. Dunning-Kruger central:

> *The Dunning-Kruger effect is a cognitive bias in which unskilled individuals suffer from illusory superiority, mistakenly rating their ability much higher than average. This bias is attributed to a metacognitive inability of the unskilled to recognize their mistakes...*

Kruger and Dunning proposed that, for a given skill, incompetent people will

- tend to overestimate their own level of skill;
- fail to recognize genuine skill in others;
- fail to recognize the extremity of their inadequacy;

- recognize and acknowledge their own previous lack of skill, if they can be trained to substantially improve.

Big if, should you ask me.

Sound familiar to anyone besides me? This is not an isolated example, although extreme and rare. Almost all of the cases I see are from outside my system, such as this example. What is it about infections where so many people have the illusion that they know what they are doing when they do not and their inability to recognize it, or even just look it up or call someone? Got me.

Rationalization

As an exercise, see how long it takes you to find a review and/or guidelines for the evaluation and treatment of S. aureus bacteremia. It should take you less than sixty seconds. Unless you are an ortho-trauma attending.

Poll Results

You are pissy. Mellow out, dude. I

- think ID is easy. Just look at the sensitivity and pick a drug. 8%
- rely on strong big gun powerful antibiotics, and so do not worry about infections. 3%
- think ID is hard. But not as hard as the Loop of Henle. 29%
- think you are an arrogant ignoramus; who are you to say? 2%
- read this book as a substitute for real knowledge. 42%
- Other Answers 15%

    -think you should rely on the advice of that cute drug rep you met at the conference last week.

    -think you're right.

Lacs

Not Lacrosse, although Lacs is a big deal at my kid's school. Lacrosse didn't even exist back in the Middle Ages when I was in high school. Lacs is a laceration.

The patient had a long and complicated course, a coronary artery bypass graft (CABG) with valve replacement, and a subsequent stroke. While in rehab she develops increasing neck pain and fever. At first, it was thought to be a muscle strain, but it increased in severity and led to a CT scan. The scan showed phlegmon and gas in the paraspinous tissues of the neck, and an MRI demonstrated discitis in the cervical spine as well, but no epidural abscess.

A debridement cleaned out the pus and the cultures grew mouth flora: *S. anginosus* and anaerobic gram-negative rods. How did those get there? Must have been an esophageal laceration from one of the other tubes down the back of the throat: transesophageal echocardiography or intubation, or both. It is rare, but I have seen a smattering of neck abscess due to medical procedures of all types. But the only way those bacteria could get into the paraspinous tissues is from a hole in the esophagus.

The literature suggests the rates of esophageal perforation from TEEs are 1 in 5,000. That seems like a lot, but I have watched enough TEEs over the years to know it is not always easy to get that big tube around the corner and down the esophagus. There are other reasons as well:

> *"A wide spectrum of causes was reported, e.g. instrumentation of the oesophagus 136 (47.6%), spontaneous rupture 89 (31.1%) or procedures otherwise related to surgical intervention 9 (3.1%). One third of the patients started conservative treatment 91 (31.9%). The majority of the patients were transferred to a thoracic surgery*

department for further treatment: about 25% of patients underwent surgery. The average hospitalization time was 18 days. The mortality rate was 21%."

The most impressive esophageal laceration I have seen was with a young man who, while drinking a wee bit too much beer, swallowed a mostly unchewed tortilla chip. Those things are like ninja throwing stars and ripped a Titanic into an iceberg rent along the esophagus.

My patient's tear sealed spontaneously and he is getting better on a long course of antibiotics.

I chew very carefully and do not expect a TEE anytime soon. But given how easily I gag, I am more at risk for aspiration pneumonia.

Rationalization

Bavalia, N. et al. Esophageal perforation, the most feared complication of TEE: early recognition by multimodality imaging. Echocardiography 28, E56-59 (2011).

Ryom, P. et al. Aetiology, treatment and mortality after oesophageal perforation in Denmark. Danish medical bulletin 58, A4267 (2011).

Meislin, H. & Kobernick, M. Corn chip laceration of the esophagus and evaluation of suspected esophageal perforation. Annals of emergency medicine 12, 455-457 (1983).

Low Levels? I Was Looking for Increased.

The patient has the abrupt onset of ankle pain, then wrist pain while having fevers and rigors.

No good reason for any of the symptoms. Just red, hot, painful, swollen joints. So she is admitted with the diagnosis of gout or septic joints. And her blood grows *Streptococcus pneumoniae*.

Odd. Septic arthritis is a complication of bacteremic pneumococcal pneumonia, but the patient did not have pneumonia, a history of joint disease, or known immunodeficiency.

Many patients with pneumococcal infected joints have underlying joint disease (especially rheumatoid arthritis) and coexistent alcoholism. Although most infections involve native joints, prosthetic joint infections comprise 13% of cases. Polyarticular disease occurs in approximately one-quarter of all patients. Most patients have a preceding or concurrent extra-articular focus of pneumococcal infection.

A spontaneous bacteremia and septic joint means I have some 'splainin to do. A lot of diseases increase the risk for invasive *Pneumococcus*, but she has none on the list.

> *There were elevated rate ratios for many of the immune-mediated diseases, for example, Addison's disease in England 3.8 (95% CI 3.4 to 4.2), autoimmune haemolytic anaemia 4.9 (4.4 to 5.3), Crohn's disease 2.2 (2.1 to 2.3), diabetes mellitus 3.7 (3.4 to 4.1), multiple sclerosis 3.7 (3.5 to 3.8), myxoedema 1.60 (1.58 to 1.63), pernicious anaemia 1.74 (1.66 to 1.83), primary biliary cirrhosis 3.3 (2.9 to 3.7), polyarteritis nodosa 5.0 (4.0 to 6.0), rheumatoid arthritis 2.47 (2.41 to 2.52), scleroderma 4.2 (3.8 to 4.7), Sjogren's syndrome 3.2 (2.9 to 3.5) and systemic lupus erythematosus 5.0 (4.6 to 5.4).*

The only risk was chemotherapy for breast cancer, but that was over a year ago. When I get an elderly pneumococcal bacteremia patient with a lack of good risks, I go looking for myeloma. And the serum protein electrophoresis (SPEP) came back showing low gammaglobulins. The IgG levels/subtypes were half normal across -the -board. Definitely not myeloma, which would raise the levels.

So she's immunodeficient. Huh. Is this her baseline? No history to suggest common variable immune deficiency. Is it from the chemotherapy? The closest I can find is an article saying that prior breast cancer chemotherapy reduces immunity to smallpox. The discussion mentions that there are no studies to suggest that breast cancer chemotherapy leads to increased infections, but notes the lack of literature on the effects of various cancer chemotherapies on immune function in adults. It is a lack I can confirm after time wandering the PubMeds.

The pus is drained, the bug dying from beta-lactams, and the patient is better, but I lack a good just -so story to tie it all together. The best I can conclude is two trues and a maybe related. Just like Lucy?

Rationalization

Raad, J. & Peacock, J.E., Jr. Septic arthritis in the adult caused by Streptococcus pneumoniae: a report of 4 cases and review of the literature. Seminars in arthritis and rheumatism 34, 559-569 (2004).

Wiser, I. et al. Immunosuppressive treatments reduce long-term immunity to smallpox among patients with breast cancer. The Journal of infectious diseases 201, 1527-1534 (2010).

Wotton, C.J. & Goldacre, M.J. Risk of invasive pneumococcal disease in people admitted to hospital with selected immune-mediated diseases: record linkage cohort analyses. Journal of epidemiology and community health 66, 1177-1181 (2012).

Nine famous TV catchphrases that were never actually said on their shows: https://www.metv.com/lists/9-famous-tv-catchphrases-that-were-never-actually-said-on-their-shows

Poll Results

Some 'splainin to do!

- is vital to the enjoyment of the practice of medicine. 57%
- is a waste of time and delays getting home. 3%
- is great as long as someone else looks it up. 0%
- changes so often they are not worth learning. 10%
- are rationalizations for true ignorance. 30%

Peripatetic Pastimes

"There is no lykelihoode that those thinges will bring gryst to the mill" *The Sermons of J. Calvin upon Deuteronomie*, 1583

WORK has been oddly slow for the last week. It is all part of the continuing and depressing decline in nosocomial infections in all of my hospitals. I never dreamed infection control could be so damn successful. While I am awaiting some testing that may (or may not) give me a denouement or two to spice up a case history, I got nothing today.

I am the sole (and the soul) ID doc at three hospitals that are geographically separated, and I spend an inordinate amount of time driving from place to place. I can easily spend two hours a day in the car, especially if the traffic is bad or if the consults that come in are poorly timed.

What makes driving tolerable are audio books. I read a book a week with all the time I spend in the car and on my daily constitutional. I have gotten to the point where I prefer audio books, as the readers do such a good job of staging the books. I also drive better as a result, since I go the speed limit to increase my reading time. That, and I lost my license a year ago for a month due to one too many speeding tickets. Public transportation is not the optimal way to commute between hospitals, I can assure you.

I don't like medically related art. There are always little errors and it is rare for the mistakes to be ignored because the underlying art is so good. *House*, the last medical TV show I watched, drives me nuts that way. Even *Doc Marten* gets on my nerves: he is a vascular surgeon having what appears to be an encyclopedic knowledge of obscure infectious and other diseases. I enjoy the character, but medical believability? Puh leaze. The shows never reach the point where I can suspend my critical thinking. The only show that gets it right most of the time is *Scrubs*.

I also avoid medically related books as a rule. But there was a sale on Audible last month, and *Doomsday Book* by Connie Willis and read by Jenny Sterling was on sale, and I had heard it was a good sci-fi book, so I gave it a shot. It is a time travel story and concerns two outbreaks: plague in 1348, and influenza in 2048. There are some annoying ID details that I have to credit to the book being published in 1992, although when a character alluded to the "Canadian Goose flu pandemic of 2010," it was almost psychic.

Large swaths of humanity infected. What more could one want? No flu or plague likely for me this week, however. It is an unfortunate fact that the pain and suffering of others is the required grist for this book's mill. So it makes me a wee bit guilty to hope for a good infection or two to come along. I guess for now it will have to be fictional infections.

Poll Results

The TV shows that captures medicine best is:

- *Scrubs*. 34%
- *ER*. 5%
- *St. Elsewhere*. 7%
- *Marcus Welby, MD*. 2%
- *Monty Python's Flying Circus*. 46%
- Other Answers 5%
  -*House*, oy vey, NOT.

UFC: Occam vs. Hickam. Round Five.

IT is always a problem in medicine. How many of the patient's symptoms can be explained by one process, or do you have to multiply the possibilities needlessly?

I am strongly of the Occam mindset, and most of the time it serves me well. This week, I am not so certain I shouldn't be in Hickam's camp. Hickam's dictum says, "The patient can have as many diseases as he damn well pleases."

The patient comes in with neutropenic fevers after TAC chemotherapy for breast cancer. The main complaints are right lower quadrant pain and diarrhea.

A test for *C. difficile,* a cause of colitis, is positive, but the CT scan is consistent with typhlitis. Usually, typhlitis is a disease of leukemics with profound neutropenia, where the colonic bacteria invade into the cecum, a form of mixed synergetic necrotizing fasciitis. Typhlitis has also been described as a problem in captive lowland gorillas, so if you ever see Magilla holding his right lower quadrant, worry that it is more than an appendix (although I do not know if gorillas have an appendix). Oddly, while horses have no gallbladder, they have an acupuncture gallbladder meridian. Go figure.

C. difficile as a cause of typhlitis has only been described in hamsters.

"TAC" therapy is docetaxel, doxorubicin, and cyclophosphamide, and so should be called ddc (coincidentally like the HIV medication), but oncologists sure do love them their brand names. Docetaxel and doxorubicin are rarely associated with *C. difficile* colitis, but to make it more complicated, they are also rarely associated with typhlitis. I am uncertain if the prior reported cases of typhlitis from TAC could have been *C. difficile.*

Typhlitis or *C. difficile* or both? Can't say one way or the other, so I have to treat for both. Hickam time. The white count is back up, and the abdominal exam is remarkably benign so time will hopefully heal the wounds.

And if your gorilla or hamster needs chemotherapy, avoid TAC.

Rationalization

Carrion, A.F. et al. Severe colitis associated with docetaxel use: A report of four cases. World journal of gastrointestinal oncology 2, 390-394 (2010).

Sundar, S. & Chan, S.Y. Cholestatic jaundice and pseudomembranous colitis following combination therapy with doxorubicin and docetaxel. Anti-cancer drugs 14, 327-329 (2003).

Cardenal, F., Montes, A., Llort, G., Segui, J. & Mesia, R. Typhlitis associated with docetaxel treatment. Journal of the National Cancer Institute 88, 1078-1079 (1996).

Ryden, E.B., Lipman, N.S., Taylor, N.S., Rose, R. & Fox, J.G. Clostridium difficile typhlitis associated with cecal mucosal hyperplasia in Syrian hamsters. Laboratory animal science 41, 553-558 (1991).

Lee, R.V. et al. Typhlitis due to Balantidium coli in captive lowland gorillas. Reviews of infectious diseases 12, 1052-1059 (1990).

Poll Results

To guide my thoughts I fall back on

- Occam. 21%
- Hickam. 0%
- Groucho. 31%
- Nietzsche. 14%
- Winnie the Pooh. 26%

Another Unconfirmed Great Diagnosis

As regular readers of my writing are probably aware, there is nothing in medicine that gives me a bigger thrill than making a diagnosis. Everyone goes into medicine and chooses their career path for different reasons. Some choices boggle my mind, but someone, I suppose, has to be an (fill in your own here). I like the challenge of figuring things out, but so often the damn cultures fail me. What should be there, what has to be there, isn't there. Or least it won't grow.

The patient notes his cat is in a tussle in the long grass with what is apparently a bird. Hard to say for sure in the foliage and the setting sun, but who wants their cat killing a bird?

Although intervening on the behalf of one bird is a wee bit too late. Cats kill 500,000,000 birds a year in the United States, accounting for 47% of bird deaths. And with no birds to feed on, the West Nile containing mosquito turns to us. It may be more than the housing crunch that leads to West Nile virus spread.

But it wasn't a bird he reached over to protect. It was a rat. And the rat bit him on the finger. Nothing happened for forty-eight hours, then illness: fevers, rigors, myalgias, mild headache, and then to the ER, where hypotension led to admission. The finger in question was fine except for some bruising. No local infection or tenosynovitis. A mild leukocytosis, but nothing else of note on the labs and off to the ICU for a day.

A rat bite and a fever. Hmmmm. Could it be rat bite fever? I sure thought so. Rat bite fever is due to *Streptobacillus moniliformis* (and *Spirillium minor* in Asia) and occasionally causes a fever with arthritis, with most of the cases in kids. I guess kids like to play with rats.

He did develop a maculopapular (more papular than macular) rash on the extensor surface of the arms, which is a rash of rat bite fever, but did not look like any image on the Googles. So I am close, ever so close, to the diagnosis but the cultures so far remain negative. Perhaps it will grow eventually;it is a finicky beast, but after four days I am not optimistic. I warned the lab what I expected them to grow, but my wanting it doesn't make it happen, as culture remained negative.

Could it have been something else? *Streptococci*, *Staphylococci*, and *Pasteurella* should have caused a focal infection in the finger with sepsis, and we do not have any plague or Tularemia in the metropolitan Portland, at least not yet.

Allergic to penicillin, he rapidly got better on doxycycline (and on more general sepsis antibiotics). I have seen two other cases of rat bite fever, both culture positive, both in people who caught and kept wild rats as pets, although the bug can be found in "real" pet rats as well.

I still think it is the diagnosis. As my father once told me, if the labs do not support the clinical diagnosis, it is the labs that are in error.

Rationalization

Suom, C. et al. Host-seeking activity and avian host preferences of mosquitoes associated with West Nile virus transmission in the northeastern U.S.A. Journal of vector ecology : journal of the Society for Vector Ecology 35, 69-74 (2010).

Reisen, W.K., Takahashi, R.M., Carroll, B.D. & Quiring, R. Delinquent mortgages, neglected swimming pools, and West Nile virus, California. Emerging infectious diseases 14, 1747-1749 (2008).

Sakalkale, R. et al. Rat-bite fever: a cautionary tale. The New Zealand medical journal 120, U2545 (2007).

Rat Bite Fever: http://www.nlm.nih.gov/medlineplus/ency/article/001348.htm

Poll Results

It boggles my mind why anyone would choose to be

- an ID doc. 4%
- an anesthesiologist. 19%
- a pathologist. 5%
- an Ob doc. 45%
- a surgeon. 8%
- Other Answers 18%
  - -a gastroenterologist.

-a dentist.
-an infection control nurse.
-a bird.
-an orthopod
-President of the United States.
-A doctor. Any kind. Why be a doctor when you can be a pharmacist?

....

Ugg

SOMETIMES? I don't know. I connect the dots, and it forms a picture, and looking at the result, I do not accept what I see. It is ugly. Yet I can see no other explanation that ties the facts together.

The patient has surgery on his wrist to repair a work injury, and it becomes infected. MSSA and *Citrobacter*. That is not the odd part. He is on well (or "not so well") water at home, and although the water tests fine, I credit the *Citrobacter* to water exposure from bathing. He completes a course of IV antibiotics for the tenosynovitis, and a week after the end of therapy has the sudden onset of severe pain in his forearm. Really severe. So off to the ER where a CT scan shows gas in the soft tissues with minimal inflammation. The patient is transferred to one of my hospitals.

The patient had one fever at Outside Hospital, but the fever is gone now. He has a normal white blood cell count, and besides a very tender forearm and pain moving the fingers, there is no rubor, no calor, and no tumor. But you can't ignore gas in tissues, so off to the operating room. There is a tenosynovitis, or at least thick material, but gram stain and cultures are negative. He still has his PICC (peripherally inserted central catheter) line in the other arm from the prior IV antibiotic course, but it looks fine, not infected.

After a day the blood cultures are growing gram-negative rods at Outside Hospital, as are the cultures from my lab. So the PICC is pulled, but the tip is negative.

Metastatic infection from the PICC? Seems odd. Bacteremia from the arm? How did the infection get there, since the gas is a good eight inches upstream from the incision site? And the cultures? *Stenotrophomonas maltophilia.* Odd squared. I can probably credit the well water again, and it does make it a line infection if one likes to play the odds:

> *The most common source of bacteriemia was an infected central catheter in 44 patients (43.1%); 17 (16.6%) were related to neutropenic sepsis; nine (8.8%) were from an abdominal source; six (5.9%) were from a respiratory source, and the source of the bacteriemia was unclear in 26 cases (25.5%). The majority (94.1%) of the patients had central venous access devices.*

and *Stenotrophomonas* does cause the occasional community-acquired infection:

> *Regarding the 77 patients with community-acquired S. maltophilia infections included in the identified case series, 45 had bacteremia, six ocular infections, five respiratory tract infections, four wound/soft tissue infections, two urinary tract infections, one conjunctivitis, one otitis, and one cellulitis; data were not reported for the remaining 12 patients.*

Stenotrophomonads are non-fermenters, so they do not make gas, at least from metabolizing carbohydrates, so could they be making gas in tissues? I didn't think to ask Microbiology today, and I can't get an answer I understand from the Googles, except to say there are no cases of gas from *Stenotrophomonas* infections reported. But every living creature makes some sort of gas; it is the end product of all metabolism, as my teenage children gleefully demonstrate. Anyone with an answer?

So I have a community-acquired *Stenotrophomonas* line infection with a metastatic gas-forming infection of the forearm with no positive cultures from the arm or from the line.

Ugg. As pleasing as the boots. Aesthetically icky.

Postscript

In the end it was suspected, but never proven, to be self-induced, and after discussing the diagnosis with the patient, it resolved.

Rationalization

Garazi, M., Singer, C., Tai, J. & Ginocchio, C.C. Bloodstream infections caused by Stenotrophomonas maltophilia: a seven-year review. The Journal of hospital infection 81, 114-118 (2012).

Falagas, M.E., Kastoris, A.C., Vouloumanou, E.K. & Dimopoulos, G. Community-acquired Stenotrophomonas maltophilia infections: a systematic review. European journal of clinical microbiology & infectious diseases : official publication of the European Society of Clinical Microbiology 28, 719-730 (2009).

Poll Results

When faced with ugly aesthetics I realize

- I have the wrong answer. 0%
- it is me, not the conclusion, that is ugly. 12%
- finding patterns are often an attempt to bring imaginary order to chaos. 64%
- beauty is in physics, not in medicine. 6%
- bauty is in the eye of the beholder and medicine is blind. 15%
- Other Answers 3%

  -Disease is bad for a person. Why shouldn't it be ugly as well?

There and Back Again

THE patient has the uneventful life of the solid citizen; then he develops fevers and rigors. He goes to the emergency room, and a CT scan reveals some plus/minus diverticulitis (inflammation of bulging pockets of bowel). I looked at the films, and I was unimpressed. The patient had no symptoms to suggest diverticulitis, although I have been impressed over the years that the innards have no idea where they are and how they feel. Negative signs and symptoms occur often enough with abdominal catastrophes that I don't let a benign belly exam necessarily sway me.

He is started on ciprofloxacin and metronidazole, but the fevers persist and he develops a progressive headache. On day three he has a spinal tap that demonstrates meningitis and gram-positive cocci, and they call me. The blood cultures remain negative.

Huh. I have had a few patients get pneumococcal meningitis on IV ciprofloxacin back in its heyday; ciprofloxacin is a marginal streptococcal antibiotic with poor penetration into the cerebrospinal fluid. So *Pneumococcus*? From diverticulitis? Nah. *Enterococcus*? Nah. No reason.

The next day the bacteria is identified as *Streptococcus salivarius*, a mouth streptococcus. The heck. His exam is negative; he has no stigmata of endocarditis or an odontogenic/sinus infection. The spinal tap was done without benefit of a prior CT scan, so another scan was done looking for an explanation more than the alleged diverticulitis: no explanation was found.

Usually, these bacteria are a complication of spinal taps, as I have mentioned in the past, but he has zero good reported risk factors:

> *The majority of cases of S. salivarius meningitis (39 of 58 cases, 67%) were associated with iatrogenic causes, usually following epidural anesthesia or spinal myelography. In addition to the case reported here, there were 11 cases related to a leak of CSF. Of these leak-related cases, five developed following head trauma, two cases were complications from a neurosurgical procedure, two were due to spontaneous dural defect, one was due to a sphenoid mucocele, and a single case was associated with chronic sinusitis and otitis media (like our patient). The remaining cases (5 of 57) were associated with possible translocation from the gastrointestinal tract. There was a single case report of translocation from the mouth from trauma, and one report associated with a sinus infection.*

Spontaneous? From the diverticulitis? *S. salivarius* is a mouth bug, not usually reported in the colon. Although the two body parts are connected, it is a long way from the mouth to the colon and back to the brain.

Neither of the explanations is satisfying, but probably the colon is the source. Thanks to antibiotics, he is all better.

Rationalization

Wilson, M. et al. Clinical and laboratory features of Streptococcus salivarius meningitis: a case report and literature review. Clinical medicine & research 10, 15-25 (2012).

Poll Results

There is no poll question

- but I prefer this answer. 16%
- I give false answers on all polls. 5%
- the statement of no poll question is a poll question.
- M-5 would self-destruct. 28%
- no matter what the question, the answer is ID consult. 21%
- when in doubt, answer all of these . 21%
- Other Answers 9%
    - -none of these
    - -I prefer original answers.

Doug Flutie Has Nothing on Me

TODAY in the ICU I overheard a nurse saying to a patient who survived an out-of-hospital cardiac arrest secondary to a large myocardial infarction and was waking up on a ventilator: "Don't worry. You're all right. You are in the ICU in the hospital."

I am sure the first two sentences did not correlate well with the last in the patient's mind. Medical interventions can be dangerous and scary despite the best of care. One example: All lines go bad if left in long enough: they break; they clot; they get infected. Plastic doesn't belong in the intravascular space.

The patient recently had both a mitral and aortic repair and had a very rocky postoperative course, with all of the supportive interventions available in the ICU. Acute renal failure ensued, and she required dialysis and a resultant dialysis catheter.

She was sent out to a skilled nursing facility, and after a week was at the dialysis center with a fever. That led to blood cultures.

Yeast. Which led to the line being pulled. Yeast on the tip. Which led to an ID consult. It was *Candida albicans*. She never had fungemia in the hospital. On the day of admission, one of her eyes became red, and there was a fungus ball in the back of the eye.

New valve repair and fungemia is a bad combination, and despite a negative transthoracic echocardiogram, I suggested a transesophageal echocardiogram. It showed a mitral valve vegetation. Expletive deleted, as the old Watergate tapes used to say.

Now what? For once I agree with the surgeons that she is a poor surgical risk; she barely survived the first operation, and she has multiple new comorbidities to make her chances of surviving another surgery even less. Usually, I pooh-pooh the surgeons when they protest against surgery due to prohibitive potential mortality. They never deliver: the patients always seem to do just fine or at least survive, but even I think it's a poor bet in this case.

Candida endocarditis is usually a surgical disease, although I have cured a pair of nosocomial tricuspid valve endocarditis in my day using Amphotericin B, and this vegetation is on the valve and not, as best we can tell, associated with the repair apparatus.

The PubMeds suggest a smattering of cures using caspofungin, so what the heck. Micafungin for eight weeks then lifetime suppressive azole. I am also intrigued by the thought of following the 1-3 beta-D-glucan test as a measure of fungal infection. It should go down with therapy. I suppose I am an optimist even trying, but nothing ventured, nothing lost. But Doug Flutie had it easy.

Rationalization

Lefort, A. et al. Diagnosis, management and outcome of Candida endocarditis. Clinical microbiology and infection : the official publication of the European Society of Clinical Microbiology and Infectious Diseases 18, E99-E109 (2012).

Rajendram, R., Alp, N.J., Mitchell, A.R., Bowler, I.C. & Forfar, J.C. Candida prosthetic valve endocarditis cured by caspofungin therapy without valve replacement. Clinical infectious diseases : an official publication of the Infectious Diseases Society of America 40, e72-74 (2005).

Sims, C.R. et al. Correlation of clinical outcomes with beta-glucan levels in patients with invasive candidiasis. Journal of clinical microbiology 50, 2104-2106 (2012).

http://en.wikipedia.org/wiki/Hail_Flutie

Starts as One Thing, Turns Out to Be Another, Looking for a Third.

TODAY we have a middle-aged male who has had intermittent fevers for six years. A couple of times a year he has the abrupt onset of fevers and severe myalgias and arthralgias, and he takes to bed, and they resolve in less than forty-eight hours. Uninsured, he never sought care for the episodes as they are self-limited.

This time the symptoms start the same, but they last longer and he develops right lower quadrant pain. An emergency room evaluation reveals a mass in the cecum and what look to be metastases in the liver and maybe in the left lower lobe of the lung.

The residents ask me as a curbside if I can think of anything that can cause fevers with a pattern like this for six years. Nothing comes to mind, so I say, "Biopsy the mass and see what the tumor is and call me if you find something odd." Some odd malignancy or other; I can't think offhand of an infection that persists with this pattern for six years.

So he is off to colonoscopy, and the mass is biopsied: it is an ulcerated mass of inflammation, and the biopsy shows. . ..

Amoeba. *Entamoeba histolytica.*

It is an ameboma mimicking colon cancer with metastases. Probably. The patient had spent five months this year in Southeast Asia and never had a case of colitis. It is one of those presentations I have read about but never expected to see. Even if I had done a history and physical instead of the curbside, I never would have thought of it. Common things remain common.

While there are case reports of *E. histolytica* mimicking colon cancer with metastases and lung cancer, it is unlikely that amoeba is the cause of the relapsing fevers. But to date, the evaluation of other causes, including malignancy, is negative.

And of interest, the stool test for ova and parasites is negative, although the biopsy is clearly *E. histolytica*. We have yet to go after the liver or lung lesions.

He is on metronidazole, and the current fevers are going away.

Are the liver and lung lesions also amoeba? Is there cancer as well? Is the cause of fevers going to be diagnosed?

I will let you know.

Postscript

It was all amoeba.

Rationalization

Fernandes, H., D'Souza, C.R., Swethadri, G.K. & Naik, C.N. Ameboma of the colon with amebic liver abscess mimicking metastatic colon cancer. Indian journal of pathology & microbiology 52, 228-230 (2009).

Yapar, A.F., Reyhan, M. & Canpolat, E.T. Interesting image. Ameboma mimicking lung cancer on FDG PET/CT. Clinical nuclear medicine 35, 55-56 (2010).

Poll Results

Curbsides

- are unreliable, but part of medicine. 7%
- are never at the curb. 14%
- are better than a formal consult for trivial problems since you do not have to rent a tux. 23%
- make me wish I could bill for phone advice like lawyers. 30%
- the bane of rounds. Just consult me, dammit. 25%

Yes, I Remember It Well

THERE are facts in ID you learn about just so you can pass the boreds. Not a typo. Diseases I never really expect to see given how rare they are, and even less likely to diagnose, except by mistake.

For example, the journal *Clinical Infectious Diseases* had a recent article entitled, "Identification of *Kudoa septempunctata* as the causative agent of novel food poisoning outbreaks in Japan by consumption of *Paralichthys olivaceus* in raw fish." Eat raw olive flounder; get a watery diarrhea. Since there were over 1,300 cases,

I fully expect that to be a bored question in 2020. I will never see a case, except as a fluke—although *Kudoa* is a myxosporea, not a fluke.

Ah, my last recertification. I'll be either dead, demented, and/or retired by 2030, so 2020 will be the last time I piss away $1,000 and countless hours for the American Board of Internal Medicine exam. Can I remember that factoid for eight more years when I cannot get my kids' names straight? It is not looking good.

It is nice to live in the era of the Googles, as I am likely to make the diagnosis not because I remember a disease, but because I remember that I should remember something, but what? and do a search. It is how I made a (presumptive) diagnosis of an *Aeromonas* soft tissue infection. I thought I remembered something and the Googles made me look good. I am more and more like Honoré Lachaille from *Gigi*, whose memory didn't agree with anyone else's.

The patient has Job's syndrome, named after the poor guy in the Bible who was covered with boils. Lately it's been renamed to "autosomal dominant hyper-IgE syndrome," which is less evocative. Patients have abnormal neutrophil chemotaxis as a result of decreased production of interferon gamma and excessive amounts of the immune protein IgE, which is the part of the immune system that responds to allergens. The lack of neutrophil function results in "cold" *S. aureus* abscesses—that is, abscesses without the usual intense inflammation.

He is diagnosed with an epidural abscess from methicillin-sensitive *S. aureus*, which is drained. He is transferred to my hospital and I am called when, after four weeks of nafcillin, he has a fever.

Is it the abscess in the neck? Not by MRI. And the fevers make it not so cold. He noted that he felt better this a.m. when, while brushing his teeth, a dental abscess ruptured with a lot of blood and pus. But is a dental abscess enough to give fevers and a C-Reactive Protein (CRP) of 238?

I still think it should the **C** **R**e **A**ctive **P**rotein. A highlight of my career was when the lecturer was writing the presentation on the board and the resident mentioned highly active antiretroviral therapy, the speaker wrote down HAART, and I said it was called "fairly active antiretroviral therapy" and he changed the H to an F without thinking. Hilarity ensued. Yes, I am that infantile.

I am not so sure. Dental problems are common with Job's, but I will admit to a bit of skepticism that a dental abscess is enough to give fevers. The infections seem too small and the organisms too wimpy. But the literature does not support my doubts, and he says he felt much better with the flow of dental pus.

I still suspect it is an incipient drug reaction; weeks 3-4 are when most patients rash out to nafcillin. I treated the dental infection and time will tell, as it always does, if the fevers were from the teeth.

Postscript

The fever resolved. No other diagnosis, so by default the fever was due to a dental abscess.

Rationalization

Kawai, T. et al. Identification of Kudoa septempunctata as the causative agent of novel food poisoning outbreaks in Japan by consumption of Paralichthys olivaceus in raw fish. Clinical infectious diseases : an official publication of the Infectious Diseases Society of America 54, 1046-1052 (2012).

Vally, H., Whittle, A., Cameron, S., Dowse, G.K. & Watson, T. Outbreak of Aeromonas hydrophila wound infections associated with mud football. Clinical infectious diseases : an official publication of the Infectious Diseases Society of America 38, 1084-1089 (2004).

Levinson, S.L. & Barondess, J.A. Occult dental infection as a cause of fever of obscure origin. The American journal of medicine 66, 463-467 (1979).

Job's: http://emedicine.medscape.com/article/886988-overview

Poll Results

My memory

- is the same as it ever was. 14%
- is limited to lyrics of songs from the 1980s. 14%
- is fading with exponential speed. 22%
- what was the question? 24%

- may be suspect, but I am psychic as I knew you would make the cheap joke of answer 4. 24%
- Other Answers 4%

  -is mostly electronic.

  -ia full of song lyrics from the 1960s. Obviously I am older than you.

Bread and Butter

I HAVE been very busy the last few weeks, but none of the cases are fascinomas: fever in the ICU without good reason, culture-negative sepsis in a dialysis patient, the 1/2 positive blood culture for S. viridans, and an AIDS patient with a fever. All worked up, all with no good dénouement. Less bread-and-butter ID, more Wonder Bread and I Can't Believe It's Not Butter. It is what makes up the bulk of my practice and never seems to be on the boreds. There need to be two choices on every test question: "Google it" and "We never figured it out."

The patient has slowly increasing back pain that started a year after surgery. An MRI was done and was consistent with discitis and osteomyelitis. Blood and biopsy cultures are negative. As is usually the case, the neurosurgeon declines an open biopsy, preferring to see if the patient responds to antibiotics.

Ick. It's probably a *Streptococcus* or a *Staphylococcus* and the patient has limited insurance, so I need to give a once-a-day therapy to get it covered. Ick again. Coming in once a day means daptomycin, and over the next three weeks nothing happens. The pain not really better, not really worse. Then the fevers begin. The urinalysis was negative, chest x-ray negative, MRI shows maybe better discitis. I check blood cultures and. . ..

Klebsiella pneumoniae. Pan-sensitive. Except, of course, to daptomycin. Three weeks on useless therapy, and expensive therapy at that. *K. pneumoniae* is a rare cause of line infection, but it is an even more unusual cause of *discitis.*

Didn't see that coming. Must be a line infection—unless it was in the disc space the whole time. But we pulled the implanted catheter that was being used to deliver the antibiotics and the tip was culture-negative.

So now what? I think I am going to go with ceftriaxone. But unsatisfying. Such is the practice of medicine.

Postscript
He slowly improved on a long course of antibiotics, intravenous and then oral.

Rationalization

Kouroussis, C. et al. Spontaneous spondylodiscitis caused by Klebsiella pneumoniae. Infection 27, 368-369 (1999).

Safdar, N. & Maki, D.G. Risk of catheter-related bloodstream infection with peripherally inserted central venous catheters used in hospitalized patients. Chest 128, 489-495 (2005).

For Want of a Nail

The practice of medicine is dangerous. I learned that lesson well as an intern, when I took care of a patient who was admitted with a heart attack. She had an IV placed in the field that developed septic thrombophlebitis, *S. aureus* bacteremia, and seeded her aortic valve which blew out and killed her. For the want of a nail indeed. It has been years since I have seen a disaster of similar magnitude.

In my other writing life, those in favor of fantasy-based medical therapies like to point to the dangers of modern medicine as a reason for their preference for SCAM: supplements, complementary and alternative medicine. To my mind, that is akin to deciding that since cars are dangerous you will rely on magic carpets for your transportation. But it is impressive how little things can rapidly snowball into catastrophe.

The patient is found down and admitted to the ICU. She has an IV placed in the field. Actually in the left forearm. Twenty-four hours after admission the site, which was changed after admit, is red, hot, and swollen. A cord runs up the arm and there is even one clotted vein across the chest. She is now septic with severe disseminated intravascular coagulation (DIC)—blood clots forming throughout the body, blocking small blood vessels, fatal in 20-50% of cases.

I would expect *Staphylococcus aureus*, since *S. aureus* loves to clot blood; it ain't referred to as coagulase-positive staph for nothin'. But no. All the cultures grow *Streptococcus pyogenes,* aka Group A *Streptococcus*.

Didn't expect that. And the PubMeds support my surprise. There are a whopping five cases reported of septic thrombophlebitis due to *S. pyogenes* in the literature, three of them Lemmier's syndrome. But I got to learn the term *phlegmasia cerulea dolens,* or "painful blue edema." I thought it was a Disney movie. I think I remember seeing Phlegmasia in college.

I wonder if the patient has either a clotting disorder or an issue with her immune system, given the rapidity with which the clot propagated with the sepsis and the severity of the DIC with no toxic shock syndrome or necrotizing infection. It is probably just—and I should have "just" in quotes —an issue of source control with all the infected clot.

Any bug can cause an infection at any time. It is why this book exists, but this is odder than most.

Rationalization

Klastersky, J., Daneau, D. & Cappel, R. Suppurative thrombophlebitis caused by Streptococcus pyogenes. Acta clinica Belgica 27, 401-404 (1972).

Nao, T. et al. Toxic shock-like syndrome resembling phlegmasia cerulea dolens. Internal medicine 38, 747-750 (1999).

Mrs. Brodie

I WILL always remember Mrs. Brodie. On the first day of medical school, the dean gave us a pep talk on what it means to be a doctor. He talked about Mrs. Brodie. Pre-HIPAA or he changed the name, Mrs. Brodie had months of decline until an astute physician, probably the dean, made the diagnosis of sprue and her health was restored. As a result, one classmate referred to our gross anatomy cadaver as "Mrs. Brodie" for the entire year. For years when I heard Brodie, I thought sprue and gross anatomy. Not so much now.

One of the odd things about long bones is how rarely they become spontaneously infected, at least in adults. There is the occasional, or not so occasional, infected fracture, but when you think of all of the bacteremias and cellulitises (celluliti?) that occur every year, you would think there would be the occasional hematogenous seeding or contiguous infection of bone. Yet it is extraordinarily rare.

The patient is young and healthy and has no risks for anything. Work and school define his existence until he gets pain in his mid-arm. It does not resolve and after a week he develops a red, hot lump. He goes to the ER where plain films show something lytic mid-radius. So it's off to surgery, and in the OR the lump is found to be pus with soft, necrotic bone. Cultures grow MSSA, so it is the other Brodie: the Brodie's abscess, an abscess of a long bone. I think this is maybe the third I can recall with my increasingly enfeebled memory.

It was named after Sir Benjamin Collins Brodie, 1st Baronet. Is the Baronet a woodwind or string? Nice that in the 1800s, musicians would do a little medical work on the side; better than being a barista.

This is not really a classic Brodie's abscess since it was hot and acute, and Brodie's typically presents as an indolent cold pain of the leg, mistaken initially as a tumor due to the lack of inflammation.

Hematogenous osteomyelitis—and this must have been hematogenous, as there was no trauma—is odd. In one series of 110 patients, only 20% were adults and only two involved the radius.

> *Totally 125 bones were affected in 110 cases: femur (60), tibia (35), humerus (12), fibula (6), radius (2), ulna (1) clavicle (2) and small bones (7). Two or more than two bones were involved in 15 cases.*

There is not a lot on the PubMeds concerning spontaneous osteomyelitis of long bones in adults. I credit the rarity of the process to the lack of blood and bruise to bone. With minimal blood supply, there is no conduit for the bacteria to reach the bone. I suspect that many spontaneous infections, especially from *Staphylococcus*, occur in the hematoma of trauma rather than in the organ itself. *Staphylococcus* always seems to occur in areas of recent injury.

There are several classifications for osteomyelitis that are no use whatsoever, at least to me. Debride the dead bone if you can and prescribe antibiotics. I have always given IV antibiotics up front but I am not so sure anymore. There is some OK data to suggest that oral antibiotics are efficacious, but they are all retrospective. In kids, they often use oral therapy, but the anatomy and blood flow in the bones of kids are different. I will stick with the tired and not-so-true of an IV beta-lactam.

Rationalization

Brodie's Abscess: http://emedicine.medscape.com/article/1248682-overview#aw2aab6b2b1aa

Kharbanda, Y. & Dhir, R.S. Natural course of hematogenous pyogenic osteomyelitis (a retrospective study of 110 cases). Journal of postgraduate medicine 37, 69-75 (1991).

Spellberg, B. & Lipsky, B.A. Systemic antibiotic therapy for chronic osteomyelitis in adults. Clinical infectious diseases : an official publication of the Infectious Diseases Society of America 54, 393-407 (2012).

Poll Results

In med school,

- we did not name our cadaver because it would be disrespectful. 25%
- we named our cadaver because to call it "it" would be disrespectful. 44%
- we learned on animals due to budget cutbacks. 0%
- we had to supply our own cadavers, but that was a long time ago. 15%
- we had a virtual cadaver on a windows machine. I think most organs look like a blue screen. 7%
- Other Answers 9%

  -We did not name our cadaver because... No idea why.

Great Pox Never Goes Away

SMALL is gone, Chicken is not what it once was thanks to vaccines, Cow and Monkey show up on occasion, but Great keeps coming back. It's called great for a reason.

When I started practice in the final decade of the last century, most of the syphilis I saw was in old women. They would have neurologic symptoms, dementia, or hearing loss, and a VDRL test would be checked, and it would come up positive. The husband, often a veteran of overseas wars in the pre-penicillin era, was long dead and the putative source. As the most syphilitic generation moved on and as penicillin became ubiquitous, it became rare to see a syphilitic elderly female. As a result, fewer graves were likely opened for the purpose of desecrating the dead.

There was always the ongoing syphilis in men who have sex with men, and I get called a couple of times a year to interpret the serology in patients with a positive VDRL. The VDRL test is a nontreponemal test: it doesn't look directly for the organisms that cause syphilis. Instead it looks for antibodies and is reported as a titer showing reactivity in increasing dilutions of sample, from 1:1 down to 1:2048 or more. Less than 1:1 is considered non-reactive; higher titers indicate more antibodies, and should fall as treatment progresses.

This decade I have seen more of the disease in heterosexuals, and I have also what I think are patients who are serofast: their tests remain positive even after treatment.

The general rule of thumb is that the titer of the VDRL should fall after therapy to become non-reactive, taking one year to become negative for primary syphilis, two years for secondary, and for tertiary can you guess? Yep. Three.

Prior studies evaluating rates of decline in non-treponemal titers after treatment for primary or secondary syphilis suggested that a fourfold decrease in VDRL titers at three months, or an eightfold drop at six months, represented the earliest possible times to ascertain treatment response.

I have seen a pair of patients whose titers are stuck. One at 1:2 five years after therapy, and another at 1:8 one and a half years after treatment.

In the first case, I do not know what the initial titer was, and it could be reacquisition of syphilis, although I doubt it at 1:2, which is a low titer. The other has been at the 1:8 and history does not suggest he as reacquired the infection, although patients are likely to either dissemble about sex, drugs, and rock 'n' roll or, more importantly, not truly know the provenance of their partners.

Interestingly, 21% of non-HIV-infected patients can be serofast after treatment, which was much higher than I would have supposed.

> *We found that serological cure was independently associated with young age, fewer sex partners in the past 6 months, earlier stage of infection, higher baseline RPR titers, and a J-H reaction after treatment. A relationship between the stage of infection and the baseline RPR titer was evident in predicting treatment response, because participants with primary syphilis had a higher proportion of serological cure, whereas 43%58% of patients with secondary or EL syphilis and baseline titers d1:32 were serofast at 6 months after treatment.*

The serologic response after treatment is more variable than I suspected and whether treatment failure, reacquisition, or that is the way things are, remains uncertain.

Given the vagaries of human reporting, it is probably reasonable to repeat a course of therapy, especially in patients who have risk behaviors and inaccessible prior labs (anonymous testing in other states).

The Great Pox remains a challenge and deserves its title.

Rationalization

Centers for Disease, C. & Prevention Primary and secondary syphilis--United States, 2003-2004. MMWR. Morbidity and mortality weekly report 55, 269-273 (2006).

Sena, A.C. et al. Predictors of serological cure and Serofast State after treatment in HIV-negative persons with early syphilis. Clinical infectious diseases : an official publication of the Infectious Diseases Society of America 53, 1092-1099 (2011).

Poll Results

I least trust the history about

- sex. 26%
- drugs. 5%
- rock 'n' roll. 3%
- pain level. 16%
- other physicians' conclusions: "He said I would never walk again," etc. 45%
- Other Answers 5%
  - -human nature.
  - -alcohol intake.

Five Months Qualifies for Sustained

THE patient has *Enterococcus faecalis* in the blood in January. He gets two weeks of IV ampicillin. In March he has *E. faecalis* in the blood. He has a negative TEE, looking for anything on his bovine aortic valve, and a negative CT scan of the abdomen and gets four weeks of ampicillin. This week? *E. faecalis* in the blood and an exquisitely painful hip.

The sine qua non of endocarditis is a sustained bacteremia, and I would suggest that five months qualifies.

His exam is otherwise negative for signs of endocarditis, and a repeat TEE is again negative, and tap of the hip is pending (it eventually grew the same *Enterococcus*), but he will get treated for prosthetic valve endocarditis no matter what. He must have an endovascular infection as well as a hip infection.

It is interesting how clinicians believe negative tests. As a rule, if the diagnostic studies do not support your clinical diagnosis, it is the diagnostic studies that are likely to be wrong. Two months of enterococcal bacteremia without a source would have been enough for me, no matter the TEE, to treat for endocarditis.

One of the curiosities in the last couple of years is that *E. faecalis* has taken a huge jump in high-level aminoglycoside resistance, from around 20% of isolates to approximately 95%. What is odd is how little aminoglycoside we use anymore, so I wonder what is driving that evolutionary change? Food or agricultural antibiotic use? Pigs may be the source of gentamicin resistance in Denmark, but Oregon is not a big pork state.

Therapy for high-level gentamicin resistance is problematic, and the guidelines suggest ampicillin and streptomycin for a six-week minimum. Great idea in an old, hard-of-hearing male with a creatinine of 1.8 mg/dL at baseline, so already some kidney dysfunction. With streptomycin, kiss those nephrons and the eighth cranial nerve goodbye.

I have little optimism for using daptomycin or linezolid, so I am going to try ampicillin and ceftriaxone. The combination has a marginally better published track record and, despite the three most dangerous words in medicine being, "in my experience," my experience to date has been uniformly successful. Two cases, so at least it is a series.

Postscript

Much to my surprise, the infection was cured. 3 for 3.

Rationalization

Gavalda, J. et al. Efficacy of ampicillin plus ceftriaxone in treatment of experimental endocarditis due to Enterococcus faecalis strains highly resistant to aminoglycosides. Antimicrobial agents and chemotherapy 43, 639-646 (1999).

Larsen, J. et al. Porcine-origin gentamicin-resistant Enterococcus faecalis in humans, Denmark. Emerging infectious diseases 16, 682-684 (2010).

Poll Results

The most dangerous words in medicine:

- in my experience. 12%
- let's try it and see what happens. 13%
- it's a big gun, strong, powerful antibiotic. 24%
- we are moving to an electronic medical record this year. 21%
- whoops. 27%
- Other Answers 3%

    -Don't worry.

    -We're going to have to do more with less and do it faster, so shut up and get to work.

First Case Ever

THERE are services I dread. I hate it when I get a call from an eye doc. Eye infections make me nervous, and I never know for sure how or where to put that extra h in ohphthalhmohlohgy. Something like that, I guess. It is like getting enough s's and i's in Mississisiippiii.

The other service I dread is obstetrics. Infections are so rare on the OB service that when they call me, I never feel comfortable with these patients, especially if still pregnant. This patient, at least, was postpartum. Don't have to worry about the newborn. A week after a normal delivery she presents with abdominal pain, fevers, and leukocytosis and the CT scan shows a uterine abscess, a collection of debris in the wall of the uterus.

That is odd. Muscle of all kinds is difficult to infect and the uterus more so. Uterine abscess, as opposed to endometritis, is rarely reported in the PubMeds. There is the classic actinomycosis associated with IUDs, postabortion gangrene, and a smattering of cases associated with rupture. But there are very few cases of pus collection in the wall, the myometrium, of the uterus.

Usually, infection of muscle occurs after trauma, and delivery looks to be a wee bit traumatic to me, but search as I might, I cannot find much in the literature.

It is 4 cemtimeters in diameter and too large to get better on its own. It will probably be mixed bacteria, or MRSA, probably the latter since MRSA is what tends to cause pyomyositis—if this is the moral equivalent.

It is pus, lots of white blood cells, and no organisms. OK. Mixed organisms. Gram negatives are often not seen against the background of pus. And it has. . . .

Fusobacterium nucleatum.

What the?!?

This bug causes septic thrombophlebitis of the internal jugular vein, aka Lemmier's, as well as peritonsillar abscess and perhaps acute appendicitis.

There are zero uterine infections in humans reported with *Fusobacterium*, although it is an issue in dairy cows and in Iraqi buffalo. Really. If I were doing due diligence and submitting this for publication, I would mention that while never reported ever before in humans and will likely never be seen again, the next time you see a uterine abscess, you should consider *Fusobacterium* in your differential. I hate it when journals do that: Never been seen before and will never be seen again, but be sure to think of it.

Drainage and antibiotics and she is getting all better.

Rationalization

Azawi, O.I., Omran, S.N. & Hadad, J.J. Clinical, bacteriological, and histopathological study of toxic puerperal metritis in Iraqi buffalo. Journal of dairy science 90, 4654-4660 (2007).

Kuppalli, K., Livorsi, D., Talati, N.J. & Osborn, M. Lemierre's syndrome due to Fusobacterium necrophorum. The Lancet. Infectious diseases 12, 808-815 (2012).

Centor, R.M. Expand the pharyngitis paradigm for adolescents and young adults. Annals of internal medicine 151, 812-815 (2009).

Swidsinski, A. et al. Acute appendicitis is characterised by local invasion with Fusobacterium nucleatum/necrophorum. Gut 60, 34-40 (2011).

Poll Results

The patient who makes me the most nervous is one referred from

- OB. 12%
- ophthalmology. 16%
- ID . 6%
- neurosurgery. 22%
- workman's Comp. 42%
- Other Answers 2%
  -chiropractic.

One Thing

YEAH. There is only one thing I trust, and that is the cultures. When a culture is positive, then, and only then, can you really have a handle on what is happening and why. Well, perhaps that is a wee bit o' hyperbole, and my wife doesn't read my books so I can exaggerate. But a positive culture does help clear muddy water.

The patient has a rocky course: GI bleed, sepsis, multiple organ system failure, the whole nine yards, but is not six feet under thanks to modern ICU care. I am still in mild awe at what the ICU can do these days. People who would have died in my day will often survive. The sepsis is presumed to be from the gallbladder, and, not being an operative candidate, she has drainage tubes placed in the gallbladder along with antibiotics to control the sepsis. And she improves.

The thing of it is, the gram stain and the cultures of the gallbladder have zip. But the blood? It grows *Streptococcus anginosus.* So they call me. *S. anginosus*, is, along with *S. constellatus* and *S. intermedius*, one of them there abscess-causing Strep. They can also be called the milleri group or anginosus group. Sounds like an investment brokerage that advertises on the Golf channel. Streptococcal classification wanders around in the literature like a drunken microbiologist; if you are uncertain as to the significance of a bug, hey, call your friendly neighborhood ID doc.

Find those bugs in the blood; look for pus to be drained.

> *"Streptococcus anginosus, the term suggested to cover a set of streptococci previously known under various names (milleri, MG, anginosus, intermedius, constellatus), is characterized by a propensity to create parenchymatous abscesses, essentially cerebral or hepatic, particularly within the terms of septicemia. These abscesses are sometimes difficult to detect due to a difficult or non-existent symptomatology. The authors report on four cases illustrating the necessity to search for them systematically by cerebral CT scan and abdominal echography or CT scan in all cases of septicemia caused by Streptococcus anginosus."*

The CT report, besides mentioning an enlarged gallbladder, suggests an early abscess in the left lobe of the liver that was not tapped when the gallbladder was accessed. At the time, evidently, it was not thought to be an abscess. Or maybe it was noticed, but considered a difficult stick, not a simple collection that would be amenable to easy drainage.

It has to be a collection of *S. anginosus*; it is often found in liver abscesses either alone or as part of mixed flora. I look at complicated abscesses like a pear. When you first see them they are hard and unripe. Wait. With time they soften, liquefy, and ripen; multiple collections will coalesce into an easily accessible gobbet of pus. The patient is clinically stable and so I am not worried about source control. In time she will have something to drain, or it will melt away.

So, for now, it's antibiotics and time. The cultures: they always tell the truth if you know how to listen.

Postscript

The next CT had a nice abscess that we drained and, while culture-negative, had lots of gram-positive cocci. Close enough.

Rationalization

Colnot, F. et al. [Parenchymatous abscess in Streptococcus anginosus (Streptococcus milleri) septicemia. Value of their systematic search, apropos of 4 cases]. La Revue de medecine interne 15, 715-719 (1994).

Poll Results

I find truth in

- cultures. 26%
- Fox news. 0%
- fortune cookies. 36%
- blogs. 10%
- I can't handle the truth, hence vodka. 26%
- Other Answers 2%
  - -the Bible.

Will Do ID for Food

Busy weekend. Six consults at four hospitals. Only one had insurance, and it was Medicare. Sigh. Probably did not even make my gas money. Good thing dinner and a show Saturday night was inexpensive; if you ever get a chance to see Emily Wells, do not pass it up.

During the weekend's activities I also noted, not for the first time, that there needs to be a course in how to talk to your doctor. When I ask if you have pets, for example, I want to know the kind of vermin you have, not the names of your pets. Knowing what you call your animals doesn't really provide me with an exposure history.

Friday, before I left for my peripatetic weekend, I talked to a patient who, after several weeks of therapy, is all better. It looks like I guessed right.

She had six weeks of leg rubor, dolor, calor, and tumor with no systemic signs. One area seemed to be a large quasi-nodule on the calf. Not an abscess, but more than a cellulitis. She had been on several courses of antibiotics with no relief, and she was sent to me.

Of course, she does the pedicure thing, and like most females, shaves her legs before the visit. I always shave my legs after the visit, because if you shave the legs it irritates the hair follicles and when you stick your legs is the warm water, the hair follicles dilate, letting whatever is in the water into the skin, then the follicle closes up when the legs cool, trapping the wee beasties. And then you get a folliculitis or worse.

But what is in the warm, swirling water? Lots of atypical *Mycobacteria*, mostly the *Mycobacterium chelonae/abscessus* group, but there are other acid-fast bacilli reported as well. Leg *Mycobacteria* infections after pedicures are a well-described clinical phenomenon, and I have seen a few this century. There was nothing to biopsy and it hardly seemed worth a scar, so I treated best-guess with clarithromycin, and I must have guessed right, as she is all better. Or maybe I didn't, since the alleged response to antibiotics when you do not have a culture is not the most reliable way to make a microbiologic diagnosis. But after having slowly progressed on cephalexin then trimethoprim-sulfamethoxazole, whatever it was went away with the clarithromycin, so perhaps I guessed right. Always take credit when the patient gets better.

Rationalization

Stout, J.E. et al. Pedicure-associated rapidly growing mycobacterial infection: an endemic disease. Clinical infectious diseases : an official publication of the Infectious Diseases Society of America 53, 787-792 (2011).

Poll Results

When taking a history I am most annoyed when

- the spouse answers the questions even when the patient is able. 18%
- the spouse answers and they proceed to squabble about the answer. 27%
- hearing the names of all the pets. 5%
- the patient is pissy about having to provide a history and wants me to read the chart instead so he can watch TV. 38%
- patients are garrulous at four o'clock on Friday. 12%

In Cold Blood

I EXPECT some illnesses to make people ill. It flabbers my gaster on occasion when patients are remarkably nontoxic from their infections.

The patient has a crush injury when his work transportation falls on him, breaking a smattering of bones. He takes to his bed for a few days, but eventually, he is brought in to the ER, of course at the insistence of his wife. I have noted before that most men would be dead by thirty if we did not have our wives/girlfriends/mothers hauling our sorry carcasses into the ER. And somehow, despite the often life-saving result of the XX intervention, the XY is surly about the whole thing.

He has the trauma pan-scan, and all of the broken bits are identified. Also discovered are a dozen nodules, up to 1.5 cm, some cavitating, throughout his lungs. They look like septic emboli, but the history suggests no antecedent infection symptoms and the patient has been noninflammatory the twenty-four hours he has been in the hospital. No fevers or leukocytosis. Whatever it is in his lungs is apparently indolent and perhaps self-limited incidentalomas given the lack of clinical findings.

So I'm thinking, well, he works with horses, maybe *Nocardia* or *Rhodococcus*. Maybe *Cryptococcus* or another fungus, but none of this really fits. Nothing by history or physical to suggest endocarditis or septic thrombophlebitis to cause septic pulmonary emboli and the patient does not look in the least ill, just broken.

So I get blood cultures and in twelve hours they are growing gram positive cocci in clusters, a wee bit too soon for a contaminant. I have this shtick I use when talking to the residents. Upon waking every day, I say, everyone one should ask themselves a simple question: how can I make Mark Crislip's life better? The world would indeed be a better place if people asked this question, and the answer is simple. Before starting antibiotics for positive blood cultures, repeat the cultures. Mark Crislip, and by extension, all ID docs, love to know if a bacteremia is sustained. So to make me happy, I repeated the blood cultures, and they were again positive for *S. aureus* in less than twelve hours.

Sustained bacteremia, rapid time to positivity, and septic pulmonary emboli all point to an endovascular infection: right-sided endocarditis or septic thrombophlebitis, although both the transthoracic echocardiogram and the ultrasound are negative. And he does not look in the least ill. I have seen the occasional cold MSSA abscesses through the years, but not a cold endocarditis, but that is the presumptive diagnosis.

Rationalization

Martinez, J.A. et al. Microbial and clinical determinants of time-to-positivity in patients with bacteraemia. Clinical microbiology and infection : the official publication of the European Society of Clinical Microbiology and Infectious Diseases 13, 709-716 (2007).

Poll Results

The world would be a better place if

- more people thought about improving Mark Crislip's day; 11%
- antibiotics were given after the blood cultures rather than before. 20%
- the x in the right-hand corner closed the screen, not the application. 4%
- ice cream lowered serum lipids. 44%
- this is the best of all possible worlds. Sigh. Give me some

Zoloft. Or at least ice cream. 18%

- Other Answers 4%
    - -All of these .
    - -There were no mothers-in-law.
    - -If more served the Lord and were kind to humanity.
    - -Patients should appear to be sick with whatever the test results say. . . unless the test results are wrong. . . .

The Curves Continue to Diverge

KNOWLEDGE increases exponentially, I learn linearly. Except for spelling. And punctuation. I have accumulated a net negative ballence of knowledge over the last fifty-five years. There is always more I do not know than I do know. Being an expert is being ignorant with style and having easy access to PubMed.

My expertise is greater in those areas where I either have an interest or it is a consistent part of my practice. There are infections that I rarely, if ever, take care of, and as a result, have less knowledge of. I have never, as examples, ever treated hepatitis C or *H. pylori,* and while I have a reasonable general knowledge of the diseases, it is not close to the depth and breadth of my understanding of *S. aureus.*

And there are numerous interesting diseases that I never see since my practice is 99% inpatient. I have one hour a day, three days a week in the outpatient clinic, which is often three hours a week too many to my mind. The people I see are nice, but the diagnosis is often known and there is a little intellectual challenge to the case. My primary goal in all I do is not to be bored, and I get bored very easily.

Occasionally a patient comes into the clinic with a problem for which I have no clue what the answer is, usually because the issue has never needed to be addressed before, at least by me.

In January, during overseas travel, the patient had acute diarrhea. The diagnosis was delayed until after his return, but he and many others in the group were eventually diagnosed with *Giar-*

dia. The patient received a course of therapy and the symptoms went away, but a repeat stool showed *Giardia*, and a course of nitazoxanide was given. Now, five months later, after two courses of therapy, a repeat stool is positive, but it has been four months since any symptoms. In fact, the patient notes he has had the best stool of his life, although that may be related to diet change.

Of course, consultant's brain kicks in first. Common variable immunodeficiency and its many variants lead to chronic *Giardia*, but the patient doesn't have symptoms of *Giardia*—he is a carrier. It is common in ID to cure the infection but not to eradicate the organism, especially if the organism is not in a sterile body site, and the bowel is not a sterile body site.

What I did not know at the time was how long people could carry *Giardia* after being treated for infection, and would I let them prepare my salad?

The answer is months, even after therapy, and maybe they could make my salad, as long as they washed their hands. Really really well.

> *In conditions in which the child's nutritional baseline is excellent, such as in day care centers in the developed world, asymptomatic carriers may not always need therapy. However, it should be noted, that asymptomatically infected children may excrete Giardia for months, carry it home to family members and thus initiate infection in the household*

Most of the literature I can find is on children where the mean duration of *Giardia* in the stool is six months. Most of the data in adults is in undernourished, developing populations, although in outbreaks asymptomatic cases can be as high as 75%.

> *A communitywide survey of the city residents revealed that the majority (76%) of G. lamblia infections occurring during the epidemic period were asymptomatic and ran a self-limited course without treatment. No significant secondary, person-to-person spread occurred.*

Outside of food handlers, it would be suggested that asymptomatic carriage poses little risk to others and can be common.

There is certainly no lack of *Giardia* in the water on hikes here in the Great Pacific Northwest. Although it is more data to support my habit of only eating deep-fried food, infections being more of a worry than cardiac disease and all foods being better when fried. You know my theory: in medicine you will die of your specialty; feel sorry for the male obstetrician.

So I think I will let sleeping *Giardia* lie.

Rationalization

Saeed, H.A. & Hamid, H.H. Bacteriological and parasitological assessment of food handlers in the Omdurman area of Sudan. Journal of microbiology, immunology, and infection = Wei mian yu gan ran za zhi 43, 70-73 (2010).

Gardner, T.B. & Hill, D.R. Treatment of giardiasis. Clinical microbiology reviews 14, 114-128 (2001).

Lopez, C.E. et al. Waterborne giardiasis: a communitywide outbreak of disease and a high rate of asymptomatic infection. American journal of epidemiology 112, 495-507 (1980).

Poll Results

Knowledge

- is less important than experience. (You will never be my HCP.) 7%
- is easily available and even more easily ignored. 33%
- is power. AC or DC is unknown. 17%
- you kids today have it easy. Just try the Index Medicus. We had to use it uphill in snow. 20%
- has made me the undisputed Trivial Pursuits champion. 20%
- Other Answers 2%

  -is overwhelming.

Failure

ID is the canary in the coalmine of medicine. When the hospital is slow, ID gets sllllooooooooooowwwwwwwww. Usually, this kind of slow occurs in August when everyone is on vacation, but our nosocomial infection rates continue to plummet and with it the total on my W2. I still have a sufficiency of things to do, but I like seeing acutely ill patients. On occasion, a physician will apologize for consulting me, and I tell them I like seeing consults. ID is fun, and I enjoy the work. Showing up at the hospital with little to do is like putting the car in neutral and pushing on the gas pedal. I need sick people. Feed me, Seymore.

Two weeks ago a patient was transferred with fever and low back pain. The worry was whether his automatic implantable cardioverter defibrillator (AICD) was infected. He had two sets of blood cultures, and on day three they grew two different coagulase-negative *Staphylococci* and a *Bacillus*. Nothing else, and he was rock stable. After a careful evaluation I decided that it was most likely contamination and if I were wrong, it would make no difference since the coagulase-negative *Staphylococci* are not likely to kill him and it is impossible to cure an AICD infection without removal, and there was no data to support infection.

I put some stock in the time to positivity, since endovascular infections usually have the blood cultures pop positive in less than twenty-four hours although this usually applies to central line infections.

> *A time to positivity of 16 hours reflects high-grade bacteremia .... A time to positivity of >20 hours indicates possible contamination with a CFU 39 of <10 and active therapy may not be required.*

The back pain made me nervous. He had a negative MRI, but gram-positive bacteremia often presents with low back pain, although it is more likely with *S. aureus*. And I go looking for a reference and cannot find it. But I have it in my brain that *S. aureus* bacteremia will commonly, like 30% commonly, present with severe low back pain with no focal back infection.

And I have this memory, imparted by my attendings as a fellow, that I also cannot support with a reference, that right-sided endocarditis has a delay in the blood cultures since the lung acts as a filter resulting in a lower level of bacteremia. It is sometimes remarkable how much I know from information I have picked up over the years, much of it dating to my training, that is unsupported by a reference.

So we stopped his antibiotics and sent him home. And you know where this is going.

He had a recurrence of both the back pain and the fevers, and 3 of 5 blood cultures are growing the same coagulase-negative *Staphylococci*, although again a delay in time to positivity: they popped positive at days 3, 4, and 5.

Back on antibiotics, the fever is gone, and the back pain has resolved. Now we are looking again for an AICD infection, and it seems like it is going to have to come out.

You have to make decisions in medicine knowing no sign, symptom, or diagnostic test has 100% sensitivity and specificity. And sometimes you are going to be wrong. But if you are going to be wrong, try to do it in a way that will not harm the patient.

Well, there is one sign that has 100% sensitivity and specificity. Anyone who uses the terms "strong," "powerful," or "big gun" in reference to antibiotics knows nothing about the treatment of infectious diseases. Zero. That you can take to the bank. Anyone who says otherwise is likely Dunning-Kruger-ing.

In the end, the AICD was removed and found to be infected. The patient did fine.

Rationalization

Kassis, C., Rangaraj, G., Jiang, Y., Hachem, R.Y. & Raad, I. Differentiating culture samples representing coagulase-negative staphylococcal bacteremia from those representing contamination by use of time-to-positivity and quantitative blood culture methods. Journal of clinical microbiology 47, 3255-3260 (2009).

Poll Results

For antibiotics I like the terms

- big gun. 14%
- unexpectedly expensive. 16%
- less filling. 22%
- appropriate. 32%
- new and improved, with retsin. 14%

A First I Did Not Want

I HAVE mentioned in the past that the ID is like birding, only interesting. I have my life list, microbes of one sort or another that I have seen. I do note as I go over these essays from years past that I am increasingly unable to remember all of the cool cases I have seen. It is a good thing I have my writing to help me keep track of the cases. If you are a young ID doc, I would suggest you start and maintain a simple database of all of your consults: just bug, organ, outcome, and medical record number. When you are an old codger like me with a failing memory, you will appreciate it. On the other hand, no one will ever be OCD enough to do it.

There are also the diseases I hope never to see. Smallpox probably tops the list, followed closely by measles. There are sporadic outbreaks of measles, but mostly in kids and, while I have had two close, but not quite, cases where measles was considered, I have never seen a case.

I had also hoped to never see a multi-drug resistant organism (MDRO) resistant to everything but colistin, but that wish was denied today. A patient from a long-term care facility had pneumonia and bacteremia with an *Acinetobacter* susceptible only to colistin. You know you are running out of treatment options when you have to use a drug classified as a detergent. Next, we will be using IV Dial soap on these resistant bacteria.

There have been problems with MDRO *Acinetobacter* in infected wounds from Iraq, and it was initially thought that *Acinetobacter* was in Iraq's soil and water.

I had wondered what ecological pressure in Iraq had led to such a resistant organism. It didn't seem like Iraq could be using enough antibiotics to drive that kind of resistance. Maybe a biologic warfare experiment gone wrong? They tried to put resistance genes in a plague bacillus and stuffed them in an *Acinetobacter* by mistake?

Then the *Acinetobacter* was traced to field hospitals:

> *Our data do support a role for environmental contamination and transmission of organisms within health care facilities. ABC organisms were isolated from patient treatment areas in all 7 field hospitals sampled. Using PFGE, we identified 5 ABC cluster groups, each of which included isolates from ⩾1 hospitalized patient that were genetically related to an environmental isolate recovered from a field hospital. Furthermore, strain typing demonstrated that clinical isolates recovered from 3 patients hospitalized at a military medical facility in the United States who had not been deployed to OIF were genetically related to an environmental isolate recovered from a field hospital in Iraq.*

And from there it was determined by molecular methods (i.e., magic) that the bug traveled from Iraq to both the UK and to the United States. The leading hypothesis for the origin of this *Acinetobacter*, and the most reasonable, is England, where inadequate infection control and hand hygiene led to the spread from Europe/United States to Iraq and back.

> *This heteroclonality and reappearance of acinetobacter in personnel participating in several military actions over the past 50 years suggest multiple sources, including local foods (also a potential source of global spread), contamination of wounds in the battlefield, and environmental spread and cross-infection in field and referral hospitals.*

Pretty impressive how these bugs can spread, easily hitching a ride with humans and with their equipment. Or even in a migrating bird.

> *We have shown that antimicrobial drug resistance genes are present in 1 of the most remote areas on Earth, the Arctic. Resistant as well as multiresistant isolates of E. coli were detected in the normal flora of Arctic birds. This finding highlights the unique nature of bacterial adaptation and the complexity of dissemination of antimicrobial drug resistance.*

I hope this doesn't mean I have to become a birder.

Rationalization

Scott, P. et al. An outbreak of multidrug-resistant Acinetobacter baumannii-calcoaceticus complex infection in the US military health care system associated with military operations in Iraq. Clinical infectious diseases : an official publication of the Infectious Diseases Society of America 44, 1577-1584 (2007).

Turton, J.F. et al. Comparison of Acinetobacter baumannii isolates from the United Kingdom and the United States that were associated with repatriated casualties of the Iraq conflict. Journal of clinical microbiology 44, 2630-2634 (2006).

Munoz-Price, L.S. & Weinstein, R.A. Acinetobacter infection. The New England journal of medicine 358, 1271-1281 (2008).

Poll Results

Excitement is defined by

- ID. 29%
- curling. 18%
- NBA finals. Go Heat. 4%
- NBA finals. Go Thunder. 6%
- birding. 17%
- Other Answers 26%

  -good beer and good cultures.
  -any hockey game.
  -wife advising you, "That was a stop sign" as you enter the intersection at 35mph so you crank the Civic HARD RIGHT and hope it works. . . .
  -watching grass grow.
  -IV gauge.
  -can't tell ya.
  -Olympic fencing.
  -cat videos.
  -gardening.
  -MROs from NH patients.

Faltering Fatalism

I used to think that infections were part of the practice of inpatient medicine. You couldn't cut people and stick in those various and sundry tubes without some of them occasionally getting infected. The bugs are too virulent and the patient comorbidities too numerous to get infection rates to zero.

The past decade has demonstrated that fatalism was mostly wrong. My hospitals have invested a great of time and resources into the multitudinous interventions that decrease hospital infections and it has been successful. Each year all kinds of hospital-acquired infections have declined, and they are real declines, not playing with the data. I have the W-2 to prove it. Back in the day, I paid the mortgage with ventilator-associated pneumonias, post-op wound infections, and central line-associated bloodstream infections (CLABSIs). No longer. I think that as long as we put in Foley catheters we will see the odd urinary tract infection, but no one gets an ID consult for uncomplicated cystitis.

Most of the surgical infections I have been seeing have been late, occurring three or four weeks out from the incision. I suspect these infections have more to do with post-op wound care at home than anything we do, or do not do, in the hospital.

I saw a post-op infection this week, a superficial abscess that grew out MSSA ten days after the discharge. Why? Well, there was Type II diabetes as a risk factor, but more important was the moderately severe psoriasis. Unknown to most, psoriatic plaques are often teaming with *S. aureus,* as are most chronic skin conditions such as eczema and atopic dermatitis. Although the one outbreak was due to Group A *Streptococcus* from a surgery tech:

> *"Further use of settling plates identified the source as a technician who entered operating rooms before but not during operations. This person had worked on all days when the settling plates were positive, disseminated group A Streptococcus to the plates, and carried the type responsible for the outbreak in psoriatic lesions on the scalp."*

I still fret about chronic skin conditions, as I tend to think of humans the way in which Charles Schultz drew Pig Pen in *Peanuts*—in a constant cloud of shedding skin and bacteria. Even if the wound gets meticulous care, I imagine the patient climbing into a bed filled with the last week's accumulation of skin and *Staphylococcus*, a dense but unseen cloud of pathogens to wallow in while sleeping.

I do not know what to do about patients with a severe skin disease who need surgery. Infections occur, but how much of this is my confirmation bias is not unclear. Certainly, Hibiclens showers in the peri-operative period would help, but Bactoban in the nose would be a waste of time given the wide distribution of *Staphylococcus*. I would wonder if a few weeks of a daily change of sheets/ towels/ pajamas, and perhaps sitting on a clean sheet on the couch until the sutures are out would help. Keep the environmental load down.

Perhaps we will be able to drive post-op wound infections down even further with standardization of wound care and beating back the *Staphylococcal* miasmas at home. And I can watch my W-2 shrink further.

Rationalization

Balci, D.D. et al. High prevalence of Staphylococcus aureus cultivation and superantigen production in patients with psoriasis. European journal of dermatology : EJD 19, 238-242 (2009).

Gomes, P.L. et al. Characteristics of Staphylococcus aureus colonization in patients with atopic dermatitis in Sri Lanka. Clinical and experimental dermatology 36, 195-200 (2011).

Mastro, T.D. et al. An outbreak of surgical-wound infections due to group A streptococcus carried on the scalp. The New England journal of medicine 323, 968-972 (1990).

Crislip's Sign

As I have far fewer years to practice ID in front of me than I do behind me, I realize I have yet to have anything named after me. I wanted to be a verb, like "to swan," when I was an intern. Or at least a noun, like Groshong. It is my own fault, I suppose, since I have neither the time nor the inclination for formal publication in the peer-reviewed literature and I have never invented anything. It is easier, and more fun, to yammer on in the blogosphere.

The patient is an elderly African-American male who, while on dialysis, has the abrupt onset of severe right inguinal pain. He goes to the ER where they think it is a hernia and they reduce the hernia, but the pain persists. He has nausea and vomiting and a fever of 102 and is admitted.

Late afternoon at the time of admission his review of systems, exam, and labs are negative except for an increased white blood cell count, a marked left shift (immature white blood cells), and chronic leg edema, right more than left. They are worried about a complication of his hernia, and a CT scan shows inguinal lymphadenopathy, presumptively a reaction to whatever is going on with the hernia and bowel. He is given one from each class of antibiotic and admitted. In the antemeridian, they call me.

Overnight his leg has gone from swollen to red, hot, tender, and swollen. He looks to have a *Streptococcal* cellulitis, and his bowel has checked out OK.

I have noticed over the years that *Streptococcal*, or what looks to be *Streptococcal*, cellulitis can be preceded by about twelve hours with severe inguinal pain, which I presume is a reactive lymphadenitis from a bit of downstream bacteremia. Then there is nausea and vomiting and the finally the rubor, dolor, calor, and tumor of the leg, often far more extensive than the usual cellulitis.

Symptomatic inguinal lymphadenitis as a sentinel sign of impending cellulitis has not, as best I can tell with the Googles and the PubMeds, been officially described. I think it is a real pattern in a subset of cellulitises/celluliti, and I have mentioned this fact twice before in the blog.

One case is "in my experience."

Two cases is "in case after case."

Three cases is "in a large series."

I now have a large series and declare it yet again to be Crislip's sign: painful lymphadenitis preceding lower extremity streptococcal cellulitis. Some fellow or resident needs to get cracking and do the real work to get me immortalized. That is probably as likely as my getting a sainthood.

Now I am off to the B 52s. Rock Lobster.

Poll Results

My goal in medicine is to

- have something named after me. 13%
- retire without being sued. 25%
- get reimbursed for what I think I am worth. 16%
- understand an EMR, any EMR. 5%
- have the joy in my field that is so evident in those who are privileged to do ID. 37%
- Other Answers 4%

  -not kill people.

Never Tried the Stuff

I HAVE a friend who has never had an apple. Ever. Not even as pie. It has become an odd point of pride, and he will probably go to his grave never eating an apple. I have yet to try raw oysters, but that is more cowardice than obstinacy. I had raw horse when I was in Japan, and keep telling myself I will try a raw oyster next time, but they look like snot on a shell, and I stick with the fried oysters instead.

I have also never prescribed Zosyn or tigecycline. I have never needed to give tigecycline, and that turns out to have been lucky since tigecycline has increased mortality compared to comparator agents.

Across randomized controlled trials, tigecycline was associated with increased mortality (risk difference [RD], 0.7%; 95% confidence interval [CI], 0.1%–1.2%; P = .01) and noncure rates (RD, 2.9%; 95% CI, 0.6%–5.2%; P = .01). Effects were not isolated to type of infection or comparator antibiotic regimen, and the impact on survival remained significant when limited to trials of approved indications (I(2)= 0%; RD, 0.6%; P = .04). A pooled analysis of the 5 trials completed by early 2005 before tigecycline was approved would have demonstrated a similar harmful effect of tigecycline on survival (I(2)= 0%; RD, 0.7%; P = .06).

Conclusions. Pooling noninferiority studies to examine survival may help ensure the safety and efficacy of new antibiotics. The association of tigecycline with excess deaths and noncure includes indications for which it is approved and marketed. Tigecycline cannot be relied on in serious infections.

But I would have eventually used it had the opportunity arose, although I am deliberately slow to use new medications. I have seen too many drugs released, kill patients, and then be withdrawn from the market. I let others make that mistake.

I have never trusted beta-lactamase inhibitor combinations. I want my antibiotics to have intrinsic resistance to degradation, and fret about the pharmacokinetics and stoichiometry of those agents. Often combination therapy is less expensive with better pharmacokinetics, so I opt for alternatives. Most patients get better most of the time no matter what you give as long as the pus is drained and the antibiotics have reasonable activity, so when I am consulted on a patient whom others have prescribed Zosyn, and I sometimes commit a (zo)sin of omission and do not change the antibiotics to my preferred treatment.

The patient is admitted with fevers and right upper quadrant pain, and a CT scan shows multiple small liver abscesses, from diverticular disease. The patient is, of course, put on Zosyn as they always are. He does not get better, so they call me. I'd go with ceftriaxone and metronidazole, but source control is the issue: drain the pus, and the patient will improve. So they drain the pus and all of the cultures grow a variety of anaerobes, including a *Bacteroides gracilis* in the blood. And the patient doesn't get better. So I suggest, once the cultures are back, change to ceftriaxone and metronidazole, my usual go-to for abdominal abscesses, and she gets better.

I have no way of knowing if she would have improved anyway. Probably. And there was a concern for a clot in the portal vein, so I was wondering about pylephlebitis—septic thrombophlebitis of the portal vein—as a reason for the slow clinical response. Against pylephlebitis is the one organism in the blood, the *B. gracilis*, is not reported as a cause of pylephlebitis. It tends to be associated with head and neck and visceral abscesses, although it may be harder to kill than other *Bacteroides*.

B. gracilis strains are generally more susceptible to all agents tested than other *Bacteroides*. Metronidazole, ticarcillin/clavulanate, chloramphenicol, and imipenem were most active. Several strains of *B. gracilis* were not killed by ampicillin/sulbactam, clindamycin, or cefoxitin. Activity was variable among strains and antimicrobial agents.

Although it is hard to say for sure reading the old literature, as the taxonomy of *B. gracilis* has been in flux for years. My lab tells me, well, maybe it is *B. gracilis*. Like all odd bugs, it is more a probability than a certainty as to what it can be called. While of little clinical significance, I await the day we can identify anaerobes with more precision using molecular methods.

In the meantime, another clinical failure, perhaps, of a beta-lactamase inhibitor drug, and more likely more fuel for my confirmation bias. I am still more likely to eat a raw oyster than give Zosyn and its relatives.

Rationalization

Baron, E.J., Ropers, G., Summanen, P. & Courcol, R.J. Bactericidal activity of selected antimicrobial agents against Bilophila wadsworthia and Bacteroides gracilis. Clinical infectious diseases : an official publication of the Infectious Diseases Society of America 16 Suppl 4, S339-343 (1993).

Finegold, S.M., Song, Y. & Liu, C. Taxonomy--General comments and update on taxonomy of Clostridia and Anaerobic cocci. Anaerobe 8, 283-285 (2002).

Johnson, C.C. et al. Bacteroides gracilis, an important anaerobic bacterial pathogen. Journal of clinical microbiology 22, 799-802 (1985).

Prasad, P., Sun, J., Danner, R.L. & Natanson, C. Excess deaths associated with tigecycline after approval based on noninferiority trials. Clinical infectious diseases : an official publication of the Infectious Diseases Society of America 54, 1699-1709 (2012).

Poll Results

I intend to, but have yet to

- prescribe Zosyn. 0%
- prescribe tigecycline. 0%
- read *War and Peace*. 17%
- eat raw horse stuffed with apple and oyster. 17%
- start a regular program of exercise and good diet. Ha. 63%
- Other Answers 2%

-e pluripus unum.

Ursine Questions

THE scariest thing about Splash Mountain is not the sudden precipitous drop at the end the ride. It is the animatronic animals leading up to the drop. Especially the bears. I think they are the only animals on the ride not to wear pants. I always wonder why, some Disney pun on bear/bare?

The patient today had been in Alaska a month ago, hunting and fishing. During that time he skinned bears, and in the process cut his finger. It more or less healed on its own. Four weeks later he hits the injury over the knuckle of his middle finger, and it becomes red, hot, swollen, and tender. He gets some oral antibiotics aimed at MRSA and then an incision and drainage. The gram stain and cultures is negative, he is continued on a variety of anti-staphylococcal drugs, and while the finger improves, the fevers do not. He is admitted with fevers and back pain.

The rest of his exam/history and review of systems is negative, and the labs are unimpressive except for 15% bands, or immature neutrophils, on the white blood cell count. That means his bone marrow has been stimulated to produce more white blood cells. The finger is red and hot and swollen but looks more post-op than actively infected. So what is it? Unrelated to the bear skinning? If it were *Staphylococcus* or *Streptococcus* it should have responded to the antibiotics and the fevers should have resolved. But if from the bear, what do Alaskan black bears have?

> *In black bears, prevalence ranged from 0% for distemper and parvovirus to 27.5% for trichinellosis and 32% for tularaemia. Antibody prevalence for brucellosis (2.5%) and tularaemia (32%) were identical for grizzly bears and black bears from the geographical area of interior Alaska.*

Toxoplasmosis are not uncommon as well. Alaskan wildlife (animal, not human, we are not talking gonorrhea and herpes here) is full of potential pathogens:

> *Antibodies to Brucella spp. were detected in sera of seven of 67 (10%) caribou (Rangifer tarandus), one of 39 (3%) moose (Alces alces), and six of 122 (5%) grizzly bears (Ursus arctos). Antibodies to Leptospira spp. were found in sera of one of 61 (2%) caribou, one of 37 (3%) moose, six of 122 (5%) grizzly bears, and one of 28 (4%) black bears (Ursus americanus). Antibodies to contagious ecthyma virus were detected in sera of seven of 17 (41%) Dall sheep (Ovis dalli) and five of 53 (10%) caribou. Antibodies to epizootic hemorrhagic disease virus were found in sera of eight of 17 (47%) Dall sheep and two of 39 (6%) moose. Infectious bovine rhinotracheitis virus antibodies were detected in sera of six of 67 (9%) caribou. Bovine viral diarrhea virus antibodies were found in sera of two of 67 (3%) caribou. Parainfluenza 3 virus antibodies were detected in sera of 14 of 21 (67%) bison (Bison bison). Antibodies to Q fever rickettsia were found in sera of 12 of 15 (80%) Dall sheep. No evidence of prior exposure to bluetongue virus was found in Dall sheep, caribou, moose, or bison sera.*

No Alaskan game sushi or tartare for me. So odds are it is Tularemia, but *Brucella* is the one with the longer incubation period of up to sixty days, the maximum incubation of Tularemia being fourteen days. There was no Faget's sign (low heart rate for the degree of fever) and no focal collections or Tularemia syndromes such as ocular-glandular.

Serologies are pending and given the acuity, likely to be negative. By the time the convalescent serologies are back. he will be done with therapy, and in the meantime, I will treat for Brucella and Tularemia.

Next time I get on Splash Mountain and I see those pantsless bears I will have a new fear, the fear of contagion: animatronic *Brucella* and *Francisella*.

Postscript

He improved with no diagnosis ever made. Sigh.

Rationalization

Chomel, B.B., Kasten, R.W., Chappuis, G., Soulier, M. & Kikuchi, Y. Serological survey of selected canine viral pathogens and zoonoses in grizzly bears (Ursus arctos horribilis) and black bears (Ursus americanus) from Alaska. Revue scientifique et technique 17, 756-766 (1998).

Zarnke, R.L. Serologic survey for selected microbial pathogens in Alaskan wildlife. Journal of wildlife diseases 19, 324-329 (1983).

Poll Results

The scariest part of Disneyland is

- Splash Mountain. 1%
- Space Mountain. 8%
- the parking lot. 26%
- the lines to *Indiana Jones*. 5%
- *It's a Small World.* That song drives. me. nuts. 59%
- Other Answers 1%

  -The fact that Americans are convinced that it's fun to spend their life savings there.

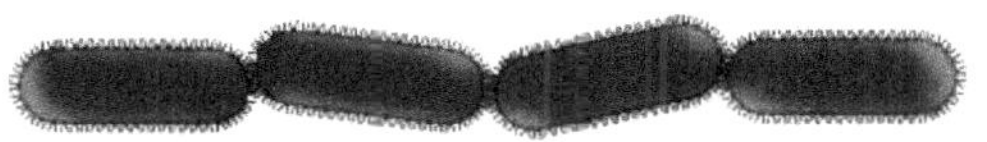

Didn't Even Consider It

THE patient comes in with twenty-four hours of progressive encephalopathy and hemiparesis. A spinal tap, which should be clear with no cells, has 1,200 red blood cells and 200 white blood cells with a 50:50 mix of lymphocytes and polymorphonuclear leukocytes. The protein level is 120 mg/100 mL, when normal is 15-60, and a glucose of 34 mg/100 mL, when normal is 50-80. After forty-eight hours they call me when there is no diagnosis.

The patient by this point is on a ventilator, and the nurse tells me he is not moving his right side. No one from the family to talk to. The chart indicates he is on steroids of an uncertain dosage for a diagnosis of some sort of bowel disease. The complete blood count is OK and blood cultures are negative at forty-eight hours.

My big worry is herpes, given the results of the spinal tap and the hemiparesis. Focal findings are common with bacterial meningitis due to alterations in blood flow, but his spinal tap does not suggest the severe inflammation that leads to focal neurologic deficits. Everything is negative, the PCR test for herpes simplex virus is pending, so I recommend at forty-eight hours with negative cultures and gram stain he needs neither isolation nor antibiotics, just the antiviral acyclovir. I ordered an MRI, which was negative, which made the diagnosis of HSV less certain. But what else? No risks that I could see for odd infections. Could be a meningovascular form of syphilis I suppose, but I am not enthusiastic. West Nile? Still none in Oregon.

I sign out for the weekend.

Next day? The spinal fluid, but not the blood, grows *Listeria* at seventy-two hours. Crap. It turns out the family is having an outbreak of gastroenteritis, which is probably *Listeria* as well. It is not an uncommon cause of gastroenteritis outbreaks.

"At least 7 outbreaks of foodborne gastroenteritis due to L. monocytogenes have been reported. Illness typically occurs 24 h after ingestion of a large inoculum of bacteria and usually lasts 2 days. Common symptoms include fever, watery diarrhea, nausea, headache, and pains in joints and muscles."

Even with the knowledge of the outbreak, I do not think I would have considered *Listeria* forty-eight hours in with negative cultures. It is the first case I have seen this century. But I should have thought of it:

"From 2001 to 2008, 1959 cases of listeriosis were reported in France (mean annual incidence 0.39 per 100,000 residents). Compared with persons <65 years with no underlying conditions, those with chronic lymphocytic leukemia had a >1000-fold increased risk of acquiring listeriosis, and those with liver cancer; myeoloproliferative disorder; multiple myeloma; acute leukemia; giant cell arteritis; dialysis; esophageal, stomach, pancreas, lung, and brain cancer; cirrhosis; organ transplantation; and pregnancy had a 100–1000-fold increased risk of listeriosis."

Fortunately, it is not the rhomboencephalitis associated with Listeria, which affects the cerebellum and brainstem and which can be fatal. He is responding rapidly to antibiotics with resolution of the hemiparesis and encephalopathy. Whew.

Listeria is a cause of meningitis that classically has few bacteria, so the gram stain is negative and there is a less impressive spinal tap, such as in this patient. When I write these entries it is often to answer a question to which I do not know the answer, and in the process of looking it up for me, you all get to learn something. The question I asked myself today was: What is the time to positivity for *Listeria* cerebrospinal fluid cultures? My Google-fu and PubMed-fu must be off, as I can't find an answer. The literature suggests a PCR will be more sensitive than cultures, but I can tell you now it should be least seventy-two hours before you consider stopping the ampicillin for suspected cases.

Rationalization

Dalton, C.B. et al. An outbreak of gastroenteritis and fever due to Listeria monocytogenes in milk. The New England journal of medicine 336, 100-105 (1997).

Ooi, S.T. & Lorber, B. Gastroenteritis due to Listeria monocytogenes. Clinical infectious diseases : an official publication of the Infectious Diseases Society of America 40, 1327-1332 (2005).

Goulet, V. et al. Incidence of listeriosis and related mortality among groups at risk of acquiring listeriosis. Clinical infectious diseases : an official publication of the Infectious Diseases Society of America 54, 652-660 (2012).

Podagra Plus

WORK is slow, and I have not had much to write about. As I enter my twenty-third year of practice, I note in passing that as a specialty, ID has been at the forefront of declining work volumes.

Compared to twenty years ago, there are fewer HIV opportunistic infections thanks to HAART, fewer neutropenic fever thanks to GCSF (granulocyte colony-stimulating factor), fewer ventilator-associated pneumonias, surgical site infections, or central line infections, thanks to the prevention bundles. The list could go on and on, at least here in Portland, of all the infections I rarely see anymore. ID as a going concern is slowly dribbling off the court. It is not just my hospital, since I was on call this weekend covering seven hospitals in two states and I had a total of three ill patients to see.

There are, at least, still some infections that need ID involvement. Gout has fooled me many times. Usually, it is a joint that looks infected but turns out to have crystals on examination of the joint fluid. I remember years ago a young female who presented with a polyarticular small joint arthritis, and I pontificated at length to housestaff that it should be parvovirus, but a tap of the joint showed uric acid crystals. At least the resident was smart enough to think beyond infection even if I wasn't.

And just how do you tell the difference between gout and infection? Gout leads to a high number of white blood cells in the joint space and erosions of the bone. So does infection. So even if you think it is gout, gout can predispose to infection, although infection is a relatively rare complication of gout compared with other arthritic conditions.

"The knee joint was the most common site of involvement, followed by the ankle, shoulder and wrist joints. Most patients had long-standing disease and subcutaneous tophi. Subcutaneous tophi rupture with secondary wound infection is the most common route of infection."

The patient this week had a tophus (a deposit of crystalline uric acid at the surface of joints) rupture at the big toe that became secondarily infected with MSSA and Group B *Streptococcus*.

Here is the odd thing, at least to me: gout, while involving any joint, has a predilection for the great toe. One would think the great toe would be the joint most likely involved with infection. Nope. It's the knee. Best I could tell, there are no reported cases of infected metatarsophalangeal joints in gout. Perhaps it is just reporting bias, but why does gout like the great toe? Yet outside of penetrating injuries/trauma, it is rare to see the great toe hematogenously infected. For gout it is hypothesized that

"A number of factors are known to reduce urate solubility and enhance nucleation of monosodium urate crystals including decreased temperature, lower pH and physical shock, all of which may be particularly relevant to crystal deposition in the foot. An association has also been proposed between monosodium urate crystal deposition and osteoarthritis, which also targets the first metatarsophalangeal joint."

One would think the same factors that lead to gout would lead to infection. Given the clinical overlap, I will be fooled again, with the same Daltryesque scream when I miss the gout. On the other hand, I always get crystals and cultures, so I have lab backup for my clinical blinders.

Rationalization

O'Connell, P.G., Milburn, B.M. & Nashel, D.J. Coexistent gout and septic arthritis: a report of two cases and literature review. Clinical and experimental rheumatology 3, 265-267 (1985).

Yu, K.H. et al. Concomitant septic and gouty arthritis--an analysis of 30 cases. Rheumatology 42, 1062-1066 (2003).

Roddy, E. Revisiting the pathogenesis of podagra: why does gout target the foot? Journal of foot and ankle research 4, 13 (2011).

Boulman, N., Slobodin, G., Sharif, D., Rozenbaum, M. & Rosner, I. Pseudopodagra: A presenting manifestation of infective endocarditis. Clinical and experimental rheumatology 23, 251-253 (2005).

Poll Results

I have a tendency to miss

- diseases that are not due to infections. 34%
- pregnancy. 22%
- diseases that are due to infections. 3%
- family history. 3%
- just about anything in the EMR. 31%
- Other Answers 6%

    -meetings.
    -repositions.

Curbsides

I DON'T mind curbsides. I work with very good hospitalists, many of whom I have known since their med student days, and I feel comfortable giving advice about patients I haven't actually seen. They usually know not to mention it in the chart, that my advice is to be considered generic to the issue at hand. I am at that point in my life, with far fewer quality days ahead than behind me, where time is more important than money. I would rather look at the culture data on the computer and recommend an antibiotic than spend the time to see yet another quad with a UTI and make it a formal consult.

But not always. Last week I had a call from a surgery resident asking how long I would treat MRSA bacteremia from a line that was pulled. Probably four weeks of vancomycin depending on co-morbidities, adequacy of source control and whether there was a worry of endocarditis. It was a surgery resident, so I added, by the way, multiple studies have demonstrated improved outcomes in *S. aureus* bacteremia with formal ID consultation.

I always feel a little guilty quoting those studies since they were done by ID docs. It is sort of like Ford salespersons touting Fords as the best cars. Evidently unconvinced, she took the advice but not the offer for consultation. You can lean a horse to water, but you can't teach them the medical literature. Dunning-Kruger Rules.

Today I get a consult: the patient has a *S. aureus* bacteremia, and is readmitted with fevers and vomiting. A CT scan shows extensive clot in the inferior vena cava and the lungs and probable septic emboli. Old echocardiograms showed a clot in the right ventricle, although that was not seen on the most recent ECHO, and the valves are not normal. The note said, "ID recommended 4 weeks of IV antibiotics."

I generally don't care about issues of liability. I do the best medicine I can, and the cards will fall will they will, but this fried my bacon. The practice of medicine would be impossible without the constant back and forth of asking each other questions. And not half as much fun. It is my fault, I know. The surgery residents rotate from the university, and I should have known better and reminded them that I am not giving specific advice for a specific patient.

But still. So I sent a gentle reminder to my colleagues that a curbside is not a consult, and while I am happy to kibitz at length on any patient, it is not the same as a careful evaluation and recommendation.

Postscript

The patient did fine and hospital policy was changed. All *S. aureus* bacteremias get an ID consult.

Rationalization

Nagao, M. et al. Close cooperation between infectious disease physicians and attending physicians can result in better management and outcome for patients with Staphylococcus aureus bacteraemia. Clinical microbiology and infection : the official publication of the European Society of Clinical Microbiology and Infectious Diseases 16, 1783-1788 (2010).

Robinson, J.O. et al. Formal infectious diseases consultation is associated with decreased mortality in Staphylococcus aureus bacteraemia. European journal of clinical microbiology & infectious diseases : official publication of the European Society of Clinical Microbiology 31, 2421-2428 (2012).

...And there are more if you go looking.

Poll Results

Curbsides are

- a good way to learn, a bad way to practice. 31%
- I like them as I can ignore the advice if I don't like it. 11%
- I always get a real consult. I like throwing the poor procedureless schmuck a bone. 2%
- are fine. I know enough ID so I don't need no stickin' smarty pants ID doc to get in my way. 3%
- only as good as the curbsider. 49%

Cockroaches

EVERYONE gets satisfaction from different aspects of medicine. At least I hope so. Some parts of medicine seem so relentlessly dull I don't see how anyone could enjoy their job, but to each their own. For me, the reason why I get out bed in the morning is the pleasure of (maybe) finding things out. I like to make a diagnosis or discover why the patient may have had their particular infection. Sometimes I come up with a just-so story, but hey, if that is the best I get, I'll take it. I may not be right, but it is always fun.

The patient comes in with a week of fevers, chills, and new red, hot, painful marble-sizebumps on the legs, with one on the arm as well. Past medical history includes poorly controlled diabetes and a past history of intravenous drug abuse—but no drug use, he tells me, for fifteen years. Exam shows the aforementioned bumps and a large metatarsal neuropathic ulcer, uninfected, but deep.

He prefers, he informs me, to not cover the ulcer and he walks around his apartment barefoot. Given the deeply tattooed dirt on his feet, I doubt he ever wears shoes. I hear no heart murmurs, but last night I went to Ramble On, a Led Zeppelin tribute band, whose lead guitarist is my kid's guitar teacher. He is one wicked good player and is a great band, but the decibels are such that a 2/6 murmur may elude me for a few days.

I am called because on admission his blood cultures grow *Serratia liquefaciens.*

Huh. Don't believe I have ever seen that in a blood culture before. So I search the PubMeds and find the following.

1) Almost every infection is due to a nosocomial outbreak associated with contaminated medical products. He has had no contact with any IV medical products for years and denies IV drug abuse for fifteen years, so that seems unlikely. And yes, I am that gullible. It is the rare patient who denies drug abuse when their life hangs in the balance.

2) There is one case of *Serratia liquefaciens* causing pustulosis. So I think the red bumps probably are due to bacteremic *Serratia.*

3) *Serratia* is one of the bacteria found on cockroaches.

> *"The following domiciliary cockroaches were collected from restaurants in five zones of Kuala Lumpur Federal Territory, Malaysia using 1L glass beaker traps baited with ground mouse-pellets: Periplaneta americana (Linnaeus) (n = 820), Periplaneta brunnea Burmeister (n = 46), Blattella germanica (Linnaeus) (n = 12504), Supella longipalpa (Fabricius) (n = 321), Symploce pallens Stephens (n = 29) and Neostylopyga rhombifolia (Stoll) (n = 5). The following bacteria were isolated from 10 cockroach specimens: Enterobacter cloacae, Klebsiella pneumoniae ssp. pneumoniae, Klebsiella pneumoniae ssp. rhinoscleromatis and Serratia liquefaciens from 5 B. germanica; Acinetobacter calcoaceticus var. anitratus, Citrobacter diversus/amalonaticus, Escherichia vul-*

neris and K.p. pneumoniae from 3 P. brunnea; and Citrobacter freundii, Enterobacter agglomerans 4, Escherichia adecarboxylate, E. vulneris, K. p. pneumonia, K. p. rhinoscleromatis and Proteus vulgeris from 2 P. americana."

The Third World worries about cockroaches as a vector for infection, and there are no documented cases of *Serratia* spread from cockroaches to humans. Years ago, I had a woman with chronic diabetic foot ulcers who liked to walk barefoot around her twenty or so cats. She became bacteremic with *Pasteurella* with no apparent source; I think it came from walking in cat muck.

While he denies cockroaches, he apparently lives in squalor. No source is found on CT, ECHO, or exam, so I postulate that walking barefoot in his less-than-hygienic environment, where I bet there are cockroaches, he had entry by the *Serratia* from the bug to the filth to the ulcer to the bloodstream. Just so.

Rationalization

Jeffery, J. et al. Domiciliary cockroaches found in restaurants in five zones of Kuala Lumpur Federal Territory, peninsular Malaysia. Tropical biomedicine 29, 180-186 (2012).

Altes, J., Usandizaga, I., Raya, C. & Forteza-Rei, J. [Pustulosis and costochondritis caused by Serratia liquefaciens in heroin addicts]. Enfermedades infecciosas y microbiologia clinica 8, 464 (1990).

Do I Believe It?

THE patient was bit by a cat on the hand. Within twenty-four hours there is spreading erythema, fevers, and pus is draining out the bite. So the patient is off to the ER, then the OR. Despite antibiotics (Unisin) the erythema spreads, and I am called.

Now cellulitis immediately after a cat bite is often *Pasteurella*, and there is some resistance to ampicillin and I have a quasi-irrational, somewhat stoichiometric distrust of Unisin, and since there is a gram negative rod growing, I change the antibiotic to ceftriaxone and say be patient.

And the next day the pus from the OR grows a pan-sensitive *E. coli* and lots of it. I call the lab, and they say yep, it's *E. coli*, no doubt. And no *Pasteurella* in the background.

So do I "believe" the *E. coli*?

For one, there are no reported cases of *E. coli* causing a cat bite infection on PubMed. Second, *E. coli* is not a cause of non-necrotizing soft tissue infections in normal people.

On the other hand (wah wah wah) *E. coli* is found in the cat colon, and cats do lick their butts, so direct injection into the skin of *E. coli* and other material is not unreasonable. If I were to scratch my backside then offer to shake your hand, most would decline, yet people think nothing of letting their pets lick them. Ick. And the Googles does suggest *E. coli* can be grown from animal bite infections, although the references are behind a paywall.

The erythema, fever, and white blood cell count are slowing returning to normal. I bet it was *Pasteurella* and the *E. coli* is a red herring, especially since the patient had been on antibiotics before the incision and drainage. It is my job to sometimes ignore the culture, and this is one of those times I am not so certain I can. I treat both.

Rationalization

Kesting, M.R., Holzle, F., Pox, C., Thurmuller, P. & Wolff, K.D. Animal bite injuries to the head: 132 cases. The British journal of oral & maxillofacial surgery 44, 235-239 (2006).

Westling, K. et al. Cat bite wound infections: a prospective clinical and microbiological study at three emergency wards in Stockholm, Sweden. The Journal of infection 53, 403-407 (2006).

Poll Results

Nothing is more filthy than

- a dog's mouth. 5%
- a cat's mouth. 14%
- a human mouth. 41%
- a human brain. 30%
- a pig. See Pulp Fiction for the discussion. 6%

- Other Answers 4%
  - -a toddler.
  - -the human mind, as the rest of the brain functions usu ally are clean enough.

Back from Vacation

BACK from vacation. It was terrible: hiking, biking, golfing, eating, and reading in eastern Oregon. Oregon really is the best place in the United States to live, or at least to vacation. But after ten days it is good to get back in the pus.

When I had left last week the last consult was a middle-aged male with a known bicuspid aortic valve who had been ill for five months—yes, five months—with fevers, chills, sweats, and a progressive failure to thrive. The physical exam revealed a diastolic murmur that even these presbycusistic (new word I just invented) ears could hear. On his fingernails were the most wonderful splinter hemorrhages I have yet to see.

The blood cultures were growing a probable *Streptococcus* at another hospital and I figured it was going to be a *S. viridans* of some sort or another. He was going to get a transesophageal echocardiogram sometime after I left town, and I suggested that he may have a ring abscess since having symptoms greater than two months is a risk for myocardial abscess, or so my memory has it. I cannot find the reference tonight; my Google-fu must still be on vacation.

I get back to find that yes indeed, he did have two myocardial abscesses, but it wasn't a *Streptococcus* after all. The blood and abscess grew *Lactobacillus paracasei*. That would be the second *Lactobacillus* endocarditis in my twenty-five-year career, and the first on a native valve.

Lactobacillus paracasei is one of the *Lactobacilli* found in probiotics and in dairy products like yogurt, both of which he intermittently consumes.

There are a total of three cases of *Lactobacillus paracasei* causing endocarditis in the PubMeds and maybe eighty reported for all *Lactobacilli* in the literature. This book is most certainly not the literature. I have always found it odd that medical journals are called literature. *Crime and Punishment*, by Fyodor Dostoyevski, *Hamlet*, by William Shakespeare, and *The Dark Tower* series by Stephen King are literature. *Clinical Infectious Diseases*, while the best journal ever, is not literature any more than the phone book. For you youngsters, back in the day before Google, we had large books full of alphabetized names and phone numbers that you searched to find a phone number. I used to look for other Crislips in the White Pages when I traveled and rarely found one; we do not have a large representation in the gene pool. Probably just as well.

While I would love to blame the yogurt or the probiotics (which are used for the ludicrous reason of immune system boosting), I can't.

> *"Although there are rare cases of bacteremia or fungemia related to the use of probiotics, epidemiologic evidence suggests no population increase in risk on the basis of usage data."*

More likely the organism is from his endogenous flora combined with very bad luck. He has a new valve and will receive a course of ampicillin and gentimicin.

Rationalization

Salvana, E.M. & Frank, M. Lactobacillus endocarditis: case report and review of cases reported since 1992. The Journal of infection 53, e5-e10 (2006).

Snydman, D.R. The safety of probiotics. Clinical infectious diseases : an official publication of the Infectious Diseases Society of America 46 Suppl 2, S104-111; discussion S144-151 (2008).

Boost Your Immune System?: http://www.sciencebasedmedicine.org/index.php/boost-your-immune-system/

Poll Results

The best place to vacation is

- Sunriver, Oregon. 8%
- Hawaii. 16%
- Paris. 16%
- anyplace that does not smell like melena. 16%
- In my mind, man. I can go anywhere. 29%
- Other Answers 16%

  -Armpit, Wyoming.

  -at my friend's beach house at the Outer Banks because he doesn't charge me rent.

  -Syria.

  -New Orleans.

Regrets and Records

I HAVE few regrets, despite an advanced age where I should have ample opportunity to look back over a life of woulda, shoulda, coulda. Maybe I have been lucky, or maybe I am an amoral sociopath, although the two are not mutually exclusive.

There are a few minor things I wish I had done differently. One is writing a simple database of every patient I had ever seen. Name, medical record number, bug, and diagnosis. After twenty-five years it would be cool. Of course, I started medicine before such a plan was practical, but to any youngsters out there, I would recommend it. Not that anyone is OCD enough to follow through.

The other is that I wish I had kept better track of some of the extremes of labs I have seen. A blood sodium over 200 mmoles per liter when 180 can kill you, a leukemoid reaction of 85,000 white blood cells per cubic millimeter, a pH of 6.8, and a temperature of 107, all in people who survived. I once did a series of case presentations where the only history was a number and the audience had to generate a differential around the number and to come up with a diagnosis. It was a fun conference.

Today's number is greater than 1:8192. So I still do not know the upper limit, but it is impressive. The patient presented with several weeks of blurred vision and the ophthalmologist noted retinitis. He (the patient) also had increased falling and on exam had a loss of vibration and proprioception in his legs. The patient is also HIV positive with a CD4 T-cell count in the low 200s—so on the edge of being susceptible to AIDS-defining infections.

That's right. Syphilis. That number is his rapid plasma reagin (RPR) test, and it's the highest RPR ever, at least for me. Not surprisingly the lumbar puncture is positive for syphilis as well, and the patient is getting two weeks of intravenous penicillin G.

It is an interesting question as to what is the proper treatment of syphilis in HIV.

The Guidelines say,

"No treatment regimens for syphilis have been demonstrated to be more effective in preventing neurosyphilis in HIV-infected patients than the syphilis regimens recommended for HIV-negative patients"

but the data is weak, as a recent meta-analysis suggests.

"The optimal antimicrobial regimen to treat syphilis in HIV-infected subjects is unknown; guideline recommendations in this population are based on little objective data."

and they note,

"...failure estimates with 18-24 MU of aqueous penicillin for the treatment of neurosyphilis were 27.3% (6.0% to 61.0%) to 27.8% (14.2% to 45.2%)."

I think the best you can say is that his low CD4 cell count makes the risk of serologic failure higher, and so he needs close follow -up. If the whopping titer is associated with an increased risk of failure, I can't find it on the interwebs.

After all these years, treatment of Lues remains uncertain.

Rationalization

Diseases Characterized by Genital, Anal, or Perianal Ulcers: https://www.cdc.gov/std/tg2015/genital-ulcers.htm

Blank, L.J., Rompalo, A.M., Erbelding, E.J., Zenilman, J.M. & Ghanem, K.G. Treatment of syphilis in HIV-infected subjects: a systematic review of the literature. Sexually transmitted infections 87, 9-16 (2011).

Poll Results

I most regret

- my choice of medical speciality. 13%
- reading the web and knowing Snape dies at the end. 17%
- eating Soylent Green. 17%
- a white dress she had on. She was carrying a white parasol. I only saw her for one second. She didn't see me at all. But I'll bet a month hasn't gone by since that I hadn't thought of that girl. 24%
- standing on a station platform in the rain with a comical look on my face because my insides had been kicked out. 13%
- Other Answers 15%

    -the day the music died.

    -not completing a residency in internal medicine before rushing into my true calling, psychiatry.

Confusobacterium

I HAVE, on occasion, been accused of being arrogant. Perhaps. I prefer to think I have a well-defined sense of right and wrong, good and bad, true and false, when it comes to medicine and infectious disease. Usually, I am called arrogant when I disagree with the patient and tell them they do not have Lyme or Morgellons. But I think it is my job to give my patients what I think they need, not what they want. And medicine has a way of slapping you down, making you feel stupid, just when you are feeling like you know what you are doing. In medicine, you NEVER really know your job and if you think you do, it as a comforting, but dangerous, delusion.

The patient has one day of sore throat and fevers. The sore throat resolves, but the fever persists, with rigors and severe malaise.

After a week of fevers, he comes to the emergency room and is admitted. Blood cultures grow *Fusobacterium necrophorum.* Of course, this has to be Lemierre's disease—aka postanginal sepsis (angina means "neck pain," by the way), aka necrobacillosis, aka septic thrombophlebitis of the internal jugular. I have seen that disease often enough, and with *F. necrophorum* being a common cause of pharyngitis, that has to be the diagnosis. However, an ultrasound and a CT scan of the neck are negative.

Hmmmm. That's annoying.

Well, the liver function tests are a touch elevated, and I have seen *Fusobacteria* causing a pair of cases of septic thrombophlebitis of the portal vein. So a CT scan and ultrasound of the abdomen show. . .nothing. A couple of liver cysts, but nothing else.

Hmmmm again. I like to understand both the what and the why of a case, and I got nothing. I'm puzzled. It seems that this is an uncommon manifestation of an uncommon disease. Always a tricky combination.

So I put the patient on metronidazole and everything improves except the fevers and liver function tests, which ever so slowly get better. But not fast enough.

The fever tasks me. It tasks me and I shall have it! I'll chase the fever 'round the liver and 'round the jugular and 'round perdition's flame before I give it up!

So after much hand-wringing I repeat the liver ultrasound. I mean, it was negative before, but this time the liver is filled with too-numerous-to-count abscesses, all marble-size. Blind pigs and truffles; I half didn't want to pursue the liver function tests. However, the abscesses were not there a week ago and they are not amenable to drainage.

Well, at least I know why the fevers were so slow to get better and the liver function tests were elevated: the liver was chockablock full of pus. Sometimes abscesses are like pears; they take time to ripen.

I still do not know why he had *Fusobacterium* everywhere. It happens. Everything happens at least once, but I still like to know why. I shall likely be disappointed. Again.

I anticipate that with a long course of metronidazole he will slowly improve.

He did.

Rationalization

Best review of *Fusobacterium* infections ever:

Riordan, T. Human infection with Fusobacterium necrophorum (Necrobacillosis), with a focus on Lemierre's syndrome. Clinical microbiology reviews 20, 622-659 (2007).

Poll Results

I'm not arrogant, I'm

- right and you are not. 23%
- more knowledgeable. 17%
- self-assured, and with good reason. 34%
- lacking in insight. 2%
- a believer that false modesty is a waste. 21%
- Other Answers 3%

    -a hoct!

    -egomaniac.

    -pressed for time.

First World / Third World

A SENTENCE that wrecks your weekend: we recommend an erase and fresh install. Arrggghhhh. It took seventy-two hours to get my baby more or less back to fighting condition. Seventy-two hours without my MacBook Pro was painful.

I know. Boo Hoo. All of my problems are First World problems. I am well aware that the worst possible day I could have is better than the best day for most people in the world, and in the history of the world, have ever had. I don't have any real problems, at least using the rest of the world as a yardstick, but it still doesn't prevent me from whining about software issues and electronic medical records and being stuck in traffic. Again, Boo Hoo.

The patient presented to the emergency room after three months in West Africa with a fever. Awkward. Sounds like he had a fever for three months. Nope. He had been diagnosed with malaria while in Africa, and the symptoms now are the same: fever of 104 and myalgias. Both thick and thin blood smears are negative, and the patient has been compulsive with doxycycline prophylaxis. The ER sends him home since everything is negative.

Next day the fever and myalgia return, and he goes back to the ER and is admitted. I get the consult: the initial smear for malaria parasites in the blood was read as negative, he has a reduced platelet count of 54,000 per microliter when above 150,000 is normal, and there is mild liver enzyme elevation (transaminitis).

I see the patient, who tells me his symptoms are just the same as in his prior case of malaria, where the diagnosis was made on clinical grounds. But I have two negative smears, albeit made several hours after the fever.

So, I am about to pontificate on all of the other causes of fever in West Africa, especially yellow fever and dengue, when the nurse sticks her head into the room to let me know that the blood smear has *Plasmodium falciparum* after all. And I was so sure it was something else.

Another smack upside the head by a Third World problem. Malaria. There's a problem to complain about.

Doxycycline is not 100% effective for prophylaxis, and around 8% of *P. falciparum* are resistant, with variable minimal inhibitory concentrations. Given only 0.4% of his red cells were parasitized, it was having some effect. I never treat malaria without the CDC site, so Malarone it is.

Rationalization

Michel, R., Bardot, S., Queyriaux, B., Boutin, J.P. & Touze, J.E. Doxycycline-chloroquine vs. doxycycline-placebo for malaria prophylaxis in nonimmune soldiers: a double-blind randomized field trial in sub-Saharan Africa. Transactions of the Royal Society of Tropical Medicine and Hygiene 104, 290-297 (2010).

Fall, B. et al. Ex vivo susceptibility of Plasmodium falciparum isolates from Dakar, Senegal, to seven standard anti-malarial drugs. Malaria journal 10, 310 (2011).

More Confirmation Bias

Two patients this week with bacteremia, one with *S. aureus* and one with Group A *Streptococcus*. What they both had in common was a myocardial infarction with their infection. Those who listen to my Puscast know one of the recurrent themes, obsessions, or bees in my bonnet is the adverse consequences of infection. Infection leads to inflammation which is prothrombotic which leads to clot which leads to strokes, heart attacks, and pulmonary embolism.

It is a remarkable literature: vascular events go up during, and after pneumonia, flu, cystitis, zoster, and even a piddly dental work. The risk can be increased for months after the infection.

> *"The rate of vascular events significantly increased in the first 4 weeks after invasive dental treatment (incidence ratio, 1.50 [95% CI, 1.09 to 2.06]) and gradually returned to the baseline rate within 6 months. The positive association remained after exclusion of persons with diabetes, hypertension, or coronary artery disease or persons with prescriptions for antiplatelet or salicylate drugs before treatment."*

The other interesting bit of epidemiology is that patients admitted with infections tend to have higher death rates after discharge than patients with similar comorbidities who are not infected, although the studies are not as comprehensive. As an example,

"Evidence suggests that pneumonia may have significant longer-term effects and that hospitalization for pneumonia is associated with higher long-term mortality than for many other major medical conditions."

Of course, many complicating factors make causality suspect, but someday I bet the streams will cross: Infections lead to increased long-term mortality due to inflammation-induced vascular events, and perhaps some sort of intervention with the clotting cascade will be of benefit.

If my "extra" money wasn't being spent on tuition, I would invest in the new coumadin replacements. As if I know anything about investing. I'll be working until I drop dead in the hospital.

It is a good thing that the concept of boosting the immune system is pure bunkum and a sign that the speaker is a goof. If you really could boost the immune system, you would be inciting an inflammatory response which would potentially increase the risk of a vascular event.

The consequences of infections are often more complex and subtle than is apparent at first glance. Yet another reason why ID is so much cooler than the rest of medicine.

Rationalization

Matthews, J.D. Invasive dental treatment and risk for vascular events. Annals of internal medicine 154, 441; author reply 442 (2011).

Mortensen, E.M. & Metersky, M.L. Long-term mortality after pneumonia. Seminars in respiratory and critical care medicine 33, 319-324 (2012).

No Longer Tired

I WAS tired this afternoon. Been working for twelve straight days, half of them covering my partner while he was on vacation. And the Olympics have been keeping me up far past my bedtime, sometimes unexpectedly. Just as I was getting ready to go to bed last night, the BMX races came on. Why, I asked my son, is this an Olympic event? They look like circus clowns. I usually feel pity for athletes when they mess up big time. Years of work can go down the drain in an instant. But when seven of the eight BMX racers all fell like bowling pins on the turn last night, I do not think I have ever laughed so hard at a sporting event. BMX kept up far too late last night, but mostly for the LOLs. My kids hate it when I use phrases like LOL.

So I am tired when I get a consult this afternoon, one of those "we don't know what is going on" consults. I just want to go home and nap. The patient has one fever, migratory arthralgias in the wrists and ankles, and new nodules on the skin that kinda look like molluscum contagiosum. He had a similar episode a year ago.

The exam is otherwise negative.

Labs show a low white blood cell count, a normal erythrocyte sedimentation rate (a measure of inflammation), and a barely elevated CRAP (C-reactive protein, another marker for inflammation), and cocaine in the urine. He is obviously jacked on cocaine, with fidgeting and pressured speech. He says the positive test is due to the lidocaine he takes for gastroesophageal reflux disease. I think not.

Suddenly, as is always the case, I am no longer tired. I have a mystery to solve, and there is nothing that wakes me up more.

The secret to most cases of diagnostic confusion is research mode. If more docs hit the Googles or the PubMeds they may not call me, so let's keep it a secret just between us. There are no risks for infections in this case, and the pattern is not that of a collagen vascular disease. As one of our cardiologists likes to say, the top three reasons for any symptoms are drugs, drugs, and drugs, both legal and not so legal.

So I stick cocaine and arthralgias into the PubMed and voilà.

A couple of years ago I had a case of recurrent neutropenia from levamisole—a dewormer of cattle, pigs, and sheep. About 70% of cocaine is cut with levamizole, and it has far more effects than neutropenia: it causes a vasculitis with nodules, fevers, arthralgias, and arthritis.

"The clinical characteristic seems to be the presence of a painful purpuric skin rash that predominantly affects the ear lobes and cheeks, often accompanied by systemic manifestations including fever, malaise, arthralgias, myalgias, and laboratory abnormalities, for example, leukopenia, neutropenia, positive ANA, ANCA, and phospholipid antibodies."

Pretty much this case, except for the lack of painful purpura, although some of the lesions have a necrotic base and nodules have been described as well.

"Levamisole has been postulated to cause various cutaneous lesions, such as nodules, drug eruptions, palpable purpura, and ulcers."

It is the best I can do today. It's not infectious; it's not rheumatologic; it's not cancer, so it must be the drugs. And now I am no longer tired.

Rationalization

Espinoza, L.R. & Perez Alamino, R. Cocaine-induced vasculitis: clinical and immunological spectrum. Current rheumatology reports 14, 532-538 (2012).

Pearson, T., Bremmer, M., Cohen, J. & Driscoll, M. Vasculopathy related to cocaine adulterated with levamisole: A review of the literature. Dermatology online journal 18, 1 (2012).

Recognizing Rheumatologic Aspects of Cocaine Abuse. These syndromes often are confused and misdiagnosis results: https://www.rheumatologynetwork.com/articles/recognizing-rheumatologic-aspects-cocaine-abuse

Poll Results

What really jolts me alert is

- coffee. 10%
- an unknown case. 12%
- a letter from a lawyer. 27%
- a 2:00 a.m. phone call. 34%
- the knowledge that I am mortal. 17%

Amateur Calcitonin

One of my mottos (I have several) is that the three most dangerous words in medicine are "in my experience." And it is certainly true for choosing an intervention or deciding whether an intervention is effective. That's why we have clinical trials, so we do not have to rely on experience. Of course, there are always the multiple unexpected comorbidities and variables that make the application of clinical trials uncertain or impossible to an individual patient. Under those circumstances, I rely on wishful thinking and wild extrapolations to choose a therapy, which is the heart of being a specialist: making stuff up with style.

Experience is valuable diagnostically, and I have found that as time passes, it is increasingly easy to make a presumptive diagnosis. Patterns of disease and their variability have become part of my genome and recognition of the likely diagnosis in a patient often takes but a few minutes.

Of course, I still have to spend the time to do a complete history and physical, as information can be revealed that changes the initial snap judgment. Not too often, but often enough I have to be careful. We have one hospitalist who is even faster than me. Dan can reach a diagnosis with fewer presented facts than anyone I have ever seen. It is fun to watch.

The weird thing about coming up with a diagnosis, the unteachable part, is that the answer bubbles up from somewhere below awareness. My explanation for how I arrived at a diagnosis is almost always an after-the-fact rationalization. It is part of the reason why I suspect free will is mostly a myth, since so much of what I think seems to occur somewhere other than consciousness.

I get a consult for an elevated white blood cell count—35,000 cells per microliter, when 11,000 is the upper limit of normal—three days after a hip replacement in a patient with known congestive heart failure.

So I know that it is likely due to rebound leukocytosis after a prolonged hypotensive episode. In my experience, that is a common cause of increased white cells in vasculopaths.

I know that all of the usual reasons have been ruled out, or else they would not be calling me. The labs show his creatinine has tripled and he shows a positive guaiac test for blood in the stool, both evidence of intra-abdominal ischemia. And sure enough, he had a prolonged episode of intraoperative hypotension.

I go through the usual song and dance and find nothing else to account for the increased white blood cells. The only fly in the ointment was a small consolidation in the base of the lung that showed up on the CT scan—far less than you would expect could drive that kind of white blood cell count.

There are a variety of papers in the PubMeds on ischemic colitis and pancreatitis following various surgeries, especially cardiac bypass surgeries. There is, as best as my Google-fu can determine, no specific papers on diffuse ischemia causing a rebound near-leukemoid reaction. In my experience, I see these rebound leukocytosis in vasculopaths who have had prolonged hypotension, from sepsis or coronary artery bypass surgery, especially when they need an intra-aortic ballon pump. I think this is the likely diagnosis.

Of interest, the procalcitonin was up: 2.4. That means infection, right? Maybe not. As best I can tell, bowel and other ischemia does elevate the procalcitonin.

> *"Procalcitonin levels were greater in the ischemia than the non-ischemia group (9.62 vs 0.30 ng/mL; P = .0001) and in the necrosis than the non-necrosis group (14.53 vs 0.32 ng/mL; P = .0001). Multivariate analysis identified procalcitonin as an independent predictor of ischemia (P = .009; odds ratio, 2.252; 95% confidence interval, 1.225-4.140) and necrosis (P = .005; odds ratio, 2.762; 95% confidence interval, 1.356-5.627).* "

Yet another reason to not order a procalcitonin. Doesn't help determine if infected or ischemic.

Rationalization

Markogiannakis, H. et al. Predictive value of procalcitonin for bowel ischemia and necrosis in bowel obstruction. Surgery 149, 394-403 (2011).

Digital Information

I REALIZED today that it has been a long time since I bothered to look at Mandell. For those of you not part of the ID *cognoscenti*, Mandell is shorthand for Mandell, Douglas, and Bennett's *Principles & Practice of Infectious Diseases*, THE ID textbook. In fact, I did not even bother to buy the last edition, and even though the electronic version is available through the hospital library, it is no longer my go-to reference. If I need to know something, I go straight to PubMed or Google. Textbooks no longer have relevance to my learning, and I note that housestaff almost always go to UpToDate. I wonder if anyone even reads Harrison's *Principles of Internal Medicine* anymore.

This realization came about with today's consult. A patient is admitted with cellulitis and has bacteremia with Group B *Streptococci*. I am asked if there is a worry about endocarditis. Patient has end-stage liver disease, and so has a risk.

> *Diabetes, neurological impairment, and cirrhosis increase risk for invasive GBS disease. Skin, soft-tissue, and osteoarticular infections, pneumonia, and urosepsis are common presentations.*

There are no stigmata of endocarditis nor any heart murmurs, and the patient responded to antibiotics as if it was an uncomplicated infection.

In one of the older versions of Mandell, there was a nice table that showed the odds that a given *Streptococcus* represented endocarditis. I think it was in the green version. The table was discontinued in subsequent editions, and I can't find a reproduction. The nice thing about the older version of Mandell, the one I memorized for my Boreds, is that I could literally open the book to the right page for the information I needed. I can't do that with the Internet, and sometimes an odd bit of information that I know is out there is beyond my Google-fu.

The best I can say now is that in over 1,500 cases of Group B strep bacteremia, 3% or 46 had endocarditis, so 97% did not have endocarditis. No clinical reason for either an echocardiogram or a long course of IV antibiotics. Unless you live in the East Coast, apparently. We had a lecturer a few months ago who noted that on the East Coast everyone seems to get an echocardiogram, at least based on Medicare data, and that in the Great Pacific Northwest almost no one gets an echocardiogram.

I do not need textbooks anymore, or paper journals (my copies of the *New England Journal of Medicine, Clinical Infectious Diseases*, and *Journal of Infectious Diseases* go straight to the recycler; I download pdfs twice a month) nor the newspaper, or books. I have gone almost entirely digital with the rare exceptions of novels that are not electronic. But if it is not digital, I will not bother with dead tree versions. But once in a blue moon, I miss paper.

Rationalization

Farley, M.M. Group B streptococcal disease in nonpregnant adults. Clinical infectious diseases : an official publication of the Infectious Diseases Society of America 33, 556-561 (2001).

Skoff, T.H. et al. Increasing burden of invasive group B streptococcal disease in nonpregnant adults, 1990-2007. Clinical infectious diseases : an official publication of the Infectious Diseases Society of America 49, 85-92 (2009).

Poll Results

I read

- all digital. 24%
- let the journals pile up and pretend I am going to read them, then recycle one a year. 31%
- I ignore a mix of digital and paper media. 19%
- the *New York Times*; it is more or less accurate. 7%
- not at all, having an ethernet jack in the head. 7%

- Other Answers 14%
  - -a mix of digital and paper.
  - -blogs.
  - -almost entirely paper - I'm old-fashioned.
  - -print journals.
  - -Up-to-Date and Medscape, geriatrics, at your fingertips on the phone.

More Global Warming?

ANYONE who spends any time with me knows that one concept I pontificate on is the idea that the terms "strong," "big gun," or "powerful" are idiotic in relation to antibiotics. What you want are *appropriate* antibiotics: drugs that kill the bugs in the infected space with the least collateral damage and cost.

So the Chief Resident told me he overheard two surgery residents talking to each other. The first was talking about how I rant about the use of "strong," "big gun," or "powerful" and that they are terms that should never be used about antibiotics, to which his colleague replies, "What do you mean? I use big gun antibiotics all the time."

Nuff said. Dunning-Kruger lives.

The consult today is a cellulitis with bacteremia. No biggy, right? You're probably thinking Group G *Streptococcus*, the most common cause of bacteremic cellulitis in some series.

However, it is *Chromobacterium violatium*. Ain't never seen that before. Most cases are from tropical regions, where it causes cellulitis and other infections. It grows in warm water, preferring to live at temperatures between 68 and 98 F. Not common water temperatures in the Northwest. Or are they?

Vibrio species have been the waterborne illnesses most associated with warming water, both in the great Pacific Northwest and in the Baltic Sea. The last several days have been exceptionally hot for Oregon (106) and I wonder if this organism is going to be a more common pathogen as the waters of the Northwest warm.

It is worth noting that the effects of global warming may increase, and the geographic distribution of this microorganism may change in the future. Several patients with C. violaceum infection went beyond the microorganisms previous territory (conﬁned between latitudes of 35 N and 35 S) .

The patient had been swimming in a warm river water around the 45th parallel, so a good exposure. Diabetes and end-stage liver disease didn't help. One of the residents wants to write up the case so I am going to encourage him to go and measure the water temperature in the pond.

The beast is resistant to the standard antibiotics given for cellulitis, probably accounting for the high mortality rate reported. It looks like the most reliable agent is a big gun, strong antibiotic like a quinolone, and the patient is doing well on ciprofloxacin.

I have said this before, but every day I get up and go off to the hospital and it is a rare day where I don't get to see something new and cool. Or in this case warm.

Rationalization

Yang, C.H. & Li, Y.H. Chromobacterium violaceum infection: a clinical review of an important but neglected infection. Journal of the Chinese Medical Association : JCMA 74, 435-441 (2011). http://www.ncbi.nlm.nih.gov/pubmed/22036134

Baker-Austin, C. et al. Emerging Vibrio risk at high latitudes in response to ocean warming. Nature Climate Change 3, 73-77 (2013). https://doi.org/10.1038/nclimate1628

Poll Results

I think of antibiotics as

- "strong,""big gun," or "powerful." 0%
- I try not to think about antibiotics. I call ID. 24%
- redundant. Everyone get Vanc and Zosyn from me. 3%
- doomed to obsolescence. 44%
- the only therapy that definitely cures. 15%
- Other Answers 15%

    -my life.

    -BLAH.

-a way to wipe out my bowel flora.
-the second best thing in medicine after Dr. Crislip (the world needs more Crislip).

Infections of the Cellu

Two of my four consults today were cellulitis. A common disease. Allow me to pontificate on cellulitis. I am not talking about potential necrotizing fasciitis or processes associated with abscesses. I am talking diffuse erythema with fevers, chills, and leukocytosis, and no odd risks like swimming in brackish dog spit or something.

The microbiology of leg cellulitis is complex:

Strep, strep, S. lugdunensis, strep, MSSA, strep, and MRSA.

That makes for potentially complex treatment regimens. In order, and resuming no allergies, my suggestion for antibiotics in acute cellulitis are

cefazolin, cefazolin,

cefazolin, and cefazolin.

With that kind of complexity, it is no wonder I see patients on vancomycin. And Zosyn. And meropenem. And cipro. And clindamycin. I recently heard tell of a surgeon treating a post-op sepsis/wound infection with aztreonam and metronidazole. Make it stop; please make it stop; it burns; it burns sooooooooo much.*

Patience. It takes three doses of antibiotics to get therapeutic, so cellulitis will often worsen for one to two days on therapy, stabilize for a day, then either recede or fade. Be patient. Don't order an MRI or CT. Don't give broad-spectrum antibiotics.

Be. Patient.

One of the infections of the cellu that I saw today was odd: he had a chronic lateral ankle ulcer that has waxed and waned for a year after a deep vein thrombosis, for which no reason was found. I found that a puzzle, and after changing his Zosyn and vancomycin to cefazolin, I started trying to figure out why he had a chronic ulcer. He told me the ulcer had been three times the current size a month ago, but that it was much better, and on exam it looked superficial with a nice scab on it. He had some water exposure, so I was half-heartedly wondering about *Mycobacterium marinum* and *Nocardia*. He mentioned in passing that he had Klinefelter's (an extra X chromosome, or XXY) but no other issues I could find. I was stumped for a why, but while writing my note I turned to a consultant's best friend, the PubMeds.

Turns out that chronic lower extremity ulcers are part and parcel of Klinefelter's, with about forty papers on the PubMeds. There have been various hypercoagulable states associated with the disease as well. So he had a good reason for both the deep vein thrombosis and the ulcer that was not infectious in etiology. Going through his records he had never mentioned the Klinefelter's to anyone else in the past, and only in passing to me. And it was key to understanding the case. Go figure.

Rationalization

Shanmugam, V.K., Tsagaris, K.C. & Attinger, C.E. Leg ulcers associated with Klinefelter's syndrome: a case report and review of the literature. International wound journal 9, 104-107 (2012).

Dissemond, J., Knab, J., Lehnen, M. & Goos, M. Increased activity of factor VIII coagulant associated with venous ulcer in a patient with Klinefelter's syndrome. Journal of the European Academy of Dermatology and Venereology : JEADV 19, 240-242 (2005).

* if you are one of my many referring docs, no, I am not referring to you. It's that other one.

Poll Results

My go-to antibiotic(s) for cellulitis is/are

- vanco and Zosyn. Screw you. 11%
- vanco alone. 10%
- meropenam. 3%
- aztreonam and metronizadole. 5%
- chloramphenicol and lincomycin. 10%
- Other Answers 60%

  -Keflex.
  -cefazolin.
  -Ancef.
  -ceftriaxone.
  -flucloxicillin.
  -clindamycin.
  -cefazolin.
  -ephadroxil
  -cephalexin.
  -whisky 45 ml prn.
  -cefazolin.
  -ten Hail Marys and go to confession every week. If that doesn't work, try hanging a three-toed frog leg from your bedroom window at the full moon.
  -cloxacillin.
  -cefazolin.

Bactrim Creep

While I know a ton about ID, compared to what I could know, I know nothing. The real purpose of this book, and the rest of my growing multimedia empire, is to educate me. Your learning is a side effect? Complication? Adverse reaction? It is amazing (or an indication of just how ignorant I really am) that every other day I find a topic to write about that almost always adds to my knowledge base.

Today after Grand Rounds an intern hit me up for the proverbial curbside.

What, she asked me, would you use for PJP prophylaxis in a patient who is allergic to Bactrim?

Dapsone, probably. What is the reason for the PJP worry?

Steroids for lupus.

Any other reason?

Nope.

Well, I said, it has been a while since I reviewed the literature, but last time I did that was not a reason for PJP prophylaxis. I am not telling you IF the patient needs medication, but WHAT to give if they did. Dapsone. And you gave me a chapter topic, as I have some research to do.

PJP prophylaxis is used in HIV when the CD4 T cell count is < 200 or so. No argument there.

And it is reasonable for other diseases where balanced against severe adverse events. PCP prophylaxis is warranted when the risk for PCP is higher than 3.5% for adults—mostly autologous bone marrow transplantation, or a solid organ transplant or a hematologic cancer.

Others?

In my first couple of years in practice I had a patient on high-dose dexamethasone for a benign, unresectable meningioma. As the dexamethasone dose was increased, he developed progressive PJP, and when we backed off the steroids he herniated. In the end, he died of PJP. I have not seen a similar case in over twenty years.

While there are case reports in the literature, as best I can tell,the incidence of PJP is so low in rheumatologic cases that it is not worth the significant risk of sulfa drugs.

> *Although there are a number of case reports in the literature, the only collagen vascular disease with an increased incidence of PJP is Wegener granulomatosis. Oral trimethoprim-sulfamethoxazole continues to be the prophylaxis of choice for PJP.*
> *There is currently no evidence to recommend PJP prophylaxis in the non-HIV/AIDS immunocompromised population. If physicians do decide to use prophylaxis, they should always weigh the benefits with the potential risks. Further studies are needed to better quantify the risks of PJP with immunosuppressive medications.*

Of course, one bad experience with PJP and risk be damned, your patients will get prophylaxis, which seems to be the guiding reason behind most PJP prophylaxis. I put more emphasis on the risks of sulfa than on the risk of PJP in these populations, but it is a judgment call.

I have seen PJP prophylaxis for steroids in COPD as well. While there is no literature to support or deny that intervention, there is a curious literature linking PJP to progression of COPD that is in its infancy, so Bactrim in that situation may be the right drug for the wrong reason.

And speak not of legal justifications. Fear of lawyers is never a reason to do the wrong thing to your patients.

Rationalization

Green, H., Paul, M., Vidal, L. & Leibovici, L. Prophylaxis of Pneumocystis pneumonia in immunocompromised non-HIV-infected patients: systematic review and meta-analysis of randomized controlled trials. Mayo Clinic proceedings 82, 1052-1059 (2007).

Grewal, P. & Brassard, A. Fact or fiction: does the non-HIV/AIDS immunosuppressed patient need Pneumocystis jiroveci pneumonia prophylaxis? An updated literature review. Journal of cutaneous medicine and surgery 13, 308-312 (2009).

Gupta, D., Zachariah, A., Roppelt, H., Patel, A.M. & Gruber, B.L. Prophylactic antibiotic usage for Pneumocystis jirovecii pneumonia in patients with systemic lupus erythematosus on cyclophosphamide: a survey of US rheumatologists and the review of literature. Journal of clinical rheumatology : practical reports on rheumatic & musculoskeletal diseases 14, 267-272 (2008).

Norris, K.A. & Morris, A. Pneumocystis infection and the pathogenesis of chronic obstructive pulmonary disease. Immunologic research 50, 175-180 (2011).

Poll Results

I could ask about Bactrim prophylaxis, but I don't really care. I prefer

- Burgerville. 12%
- McDonald's. 26%
- White Castle. 12%
- Carl's Jr.'s.. 3%
- Jack in the Box. 9%
- Other Answers 38%
    - -Sashimi.
    - -In & Out.
    - -Five Guys.
    - -Salmon Ella's Chicken Shack, Turtle Creek.
    - -NYC Pizza.
    - -Checkers.
    - -Kippered Herring.
    - -Steak n' Shake.
    - -Wendy's.
    - -Red Robin, then Baskin-Robbins. Oh yes.

My Archenemy Surprises Me Again

If I were a superhero my archnemesis would be *S. aureus*. Few pathogens have the virulence and variability to cause disease as *S. aureus*. I used to say that *S. aureus* pays my mortgage, but so many patients with *S. aureus* infections lack insurance that is no longer the case. Why pay for health care when you can buy heroin instead?

The patient is an IV drug abuser who misses his mark and comes in with an acute infection of the thigh. More than cellulitis, in the OR it is found to be a necrotizing myositis/fasciitis and is debrided. They call me.

I fully expect a MRSA since many MRSA make the Panton-Valentine leukocidin (PVL), which is to human tissue what water is to Elphaba. This decade has seen many articles on myositis/fasciitis due to MRSA. But much to my surprise it is aMSSA. I can't test the bug to see if it makes a PVL, but it probably does. That or any of a number of tissue-destroying toxins and enzymes. I once found a jaw-droppingly long list of all of the virulence factors of *S. aureus* and I wish I had bookmarked it, as I have never been able to locate it again.

The patient was changed to a beta-lactam and is getting better. Fortunately, the blood cultures are negative so I will not have to worry about setting him up for a course of intravenous antibiotics.

I received a letter from another local health care system letting me know that they would not accept active IV drug abusers with central lines for home therapy—that the treatments would have to be once a day through a peripheral, and that the patients would be fired if they were found to be accessing their IVs for injecting drugs.

I have to admit, I no longer care (well I do, but you know) if IV drug abusers use their lines. I mostly care about curing their infections, since I have noticed that dead people can't get clean and sober. Death from infection seems a wee bit of an excessive price to pay for your addiction. We continue to treat emphysema and lung cancer in smokers and diabetes in the recalcitrant obese.

Why the stigma with IV drug abuse? Just because it is illegal? Puh-leaze. I guess that is why their motto is, "A caring difference you can feel."

Rationalization

Gerard, D. et al. Facial necrotizing fasciitis in an infant caused by a five toxin-secreting methicillin-susceptible Staphylococcus aureus. Intensive care medicine 35, 1145-1146 (2009).

Akpaka, P.E. et al. Methicillin sensitive Staphylococcus aureus producing Panton-Valentine leukocidin toxin in Trinidad & Tobago: a case report. Journal of medical case reports 5, 157 (2011).

Poll Results

When I send an IV drug abuser home

- I don't care; they brought it on themselves. It is better they die of infection than continue to use drugs. 3%
- I encourage them to not use their line and hope for the best. 10%
- I don't. I keep them in the hospital until they are cured. 9%
- I put their life and infection ahead of their IVDA, and often it works. 26%
- I do the best I can with the patient I have. 49%
- Other Answers 2%

  -Keep a close eye on my IV drugs when they're in the room.

Forgetful or Atypical

I HAVE to read the textbooks and the journals. It is how I learn about diseases and how they present. I have mentioned in the past that shortly after learning about a new disease, I see it. The one version of the Secret that may be true. Maybe.

The problem is that shortly after seeing a case I promptly forget about it. And then the patient doesn't read the textbook and presents atypically with their disease. Bad combination for a patient: forgetful doctor and odd presentation.

The patient is being treated at a skilled nursing facility for osteomyelitis with MRSA. Because of allergies she is being treated with daptomycin, and presents after four weeks with shortness of breath and diffuse bilateral lung infiltrates. Looks like an atypical pneumonia on chest x-ray, but the lack of fevers and productive cough mediate against the diagnosis. I talk to the patient and find no risks for any odd infection. I didn't have any good ideas.

Fortunately, one of the housestaff had more on the ball than me, and one of them suggested "daptomycin lung."

Of course. I had a case about a year ago and even wrote about it. And promptly forgot about it.

In this case, there was no peripheral eosinophilia and only 25% eosinophils on the blood count, and no fever. Daptomycin is supposed to cause EOSINOPHILIC pneumonitis.

> *"Definite cases had concurrent exposure to daptomycin, fever, dyspnoea with increased oxygen requirement or required mechanical ventilation, new infiltrates on chest imaging, bronchoalveolar lavage with >25% eosinophils and clinical improvement following daptomycin withdrawal."*

Close, but the patient failed, again, to be up-to-date on the literature.

Maybe it is an atypical presentation or perhaps the wrong diagnosis, but she improved rapidly with steroids and stopping the daptomycin. But most diseases are self-limited, and we never had a lung biopsy. Now I wonder what forgotten side effect to linezolid she is going to get next.

Rationalization

Kim, P.W. et al. Eosinophilic pneumonia in patients treated with daptomycin: review of the literature and US FDA adverse event reporting system reports. Drug safety 35, 447-457 (2012).

www.ingramcontent.com/pod-product-compliance
Lightning Source LLC
Chambersburg PA
CBHW071634200326
41519CB00012BA/2297

9781685530198